# STEM CELLS
*Promise and Reality*

# STEM CELLS
## *Promise and Reality*

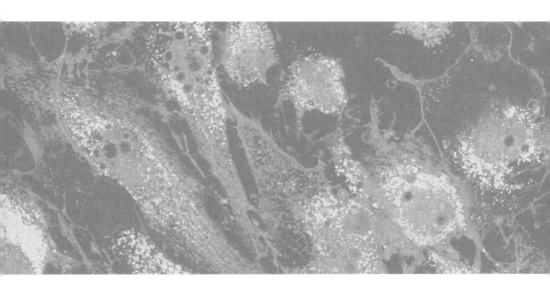

## Lygia V Pereira
University of São Paulo, Brazil

**World Scientific**

NEW JERSEY · LONDON · SINGAPORE · BEIJING · SHANGHAI · HONG KONG · TAIPEI · CHENNAI · TOKYO

*Published by*

World Scientific Publishing Co. Pte. Ltd.

5 Toh Tuck Link, Singapore 596224

*USA office:* 27 Warren Street, Suite 401-402, Hackensack, NJ 07601

*UK office:* 57 Shelton Street, Covent Garden, London WC2H 9HE

**Library of Congress Cataloging-in-Publication Data**

Names: Pereira, Lygia da Veiga, author.

Title: Stem cells : promise and reality / Lygia V. Pereira, University of São Paulo, Brazil.

Description: New Jersey : World Scientific, 2016. |
   Includes bibliographical references and index.

Identifiers: LCCN 2016035652| ISBN 9789813100183 (hardcover : alk. paper) |
   ISBN 9813100184 (hardcover : alk. paper) | ISBN 9789813100190 (pbk. : alk. paper) |
   ISBN 9813100192 (pbk. : alk. paper)

Subjects: LCSH: Stem cells. | Stem cells--Research. | Stem cells--Moral and ethical aspects.

Classification: LCC QH588.S83 P46 2016 | DDC 616.02/774--dc23

LC record available at https://lccn.loc.gov/2016035652

**British Library Cataloguing-in-Publication Data**

A catalogue record for this book is available from the British Library.

Printed in Singapore

# Contents

# Introduction — In Search of Eternal Life

Over the course of our lives, organs and tissues lose function either from disease or from the naturally occurring process of aging, requiring that we replace them, rather like changing broken car parts. Present-day medicine tries to solve this problem by **organ transplantation**, which has saved many lives, but is subject to major limitations. The first limitation is that some organs or tissues, such as nerve and muscle, we just cannot transplant or only in a very limited fashion. The second limitation is the short supply of organs and tissues for transplanting; at present, these usually come from human donors and meet only 5–10% of the demand — that is, the needs of up to 95% of patients requiring body tissue replacement or renewal go unmet.

Therefore, there is a need for new sources of organs and tissues for transplants, and science is working to develop a number of different strategies. One of these is to build **artificial organs** — bringing engineering together with the biomedical sciences in an endeavor to create machines able to restore organ functions. With the heart, basically a pump that sends blood circulating round the body, that strategy has had some success: there are now artificial hearts capable of keeping a person alive for several months. It has been possible to reproduce kidney function, although hemodialysis is extremely uncomfortable, with patients spending hours, several times a week, hooked up to machines that filter their blood. It is very unlikely, however, that organs with complex biochemical functions, such as the liver or pancreas, can be reproduced mechanically.

Another strategy is to use **animal organs**, especially from pigs, whose organs are similar to ours in size and physiology. Pigskin, for instance, can be used to treat third-degree burns, and pig pancreas cells, to treat diabetes in humans. Nonetheless, not only are many patients psychologically uncomfortable at having cells or tissues from another animal in their body, another obstacle to using xenotransplants (xeno, in Greek, means 'foreign') is the rejection of transplanted cells.

The **immune system** consists of blood cells that protect the body from outside agents, such as viruses and bacteria. However, the immune system may recognize a transplanted organ as an enemy and attack it. In order to prevent it, before a transplant is performed, donor and recipient have to be checked to see if they are immunologically similar or compatible. It is hard enough, in organ transplants, to find compatible donors whose organs that the recipient's immune system will not find "odd"; just imagine transplanting an organ from a whole other species of animal.

In order to prevent rejection of xenotransplants, some groups have genetically modified pigs so as to be compatible with humans; such animals would serve as universal organ donors. Pigs such as this are largely "invisible" to our immune system — and studies are now under way to determine whether it is feasible to transplant hearts from such animals into monkeys and eventually humans.

However, even once the psychological and (even more serious) immunological problems are solved, one other important risk remains. Such intimate contact between pig organs and human bodies may make it easier for viruses that originally infect only pigs to mutate into viruses that also attack humans. In fact, a similar scenario is the most widely-accepted hypothesis for the origin of the HIV virus, which causes AIDS. There is another very similar virus, called SIV, which normally infects only monkeys. HIV is thought to have evolved from SIV in African human populations that had come into contact with SIV infected monkey blood, and to have gained the ability to infect and multiply in humans. Accordingly, while pig organ transplants may potentially solve the problem of short supply of organs, it has the potential of creating a new biosafety risk, which, if present, will have to be recognized and controlled.

Lastly, another strategy to solve the problem of demand for organs and tissues for transplantation is what is called **tissue engineering**. Any organ is a complex system made up of different types of tissues and cells organized in a specific way for it to function properly. Instead of trying to build a complete organ in a laboratory, tissue engineering proposes to isolate *in vitro* human cells from different organs and tissues, which — when transplanted — will lead those organs to regenerate in the patient. This is where stem cells come in: these are 'wild card' cells able to multiply and develop into — among other things — neurons, muscle, liver, pancreas or blood cells etc., and so treat various different human diseases.

This book describes the various different types of stem cell, their advantages and limitations, and how they can be used for therapy and research. For those who would like to read the original scientific articles describing key points in this field, references have been included in the text. Also, to complement the illustrations, the book displays icons indicating videos that show cells in action.

Before getting into stem cells in depth, however, we will show how all these cells are organized to form a person.

# Acknowledgments

I am grateful to Dr Peter L. Pearson for critical reading of the manuscript, and to Paulo Manzi for permission to publish all the figures.

# 1 Development of the Embryo

The development of a new human being starts with the **fertilization** of an ovum (egg cell) by a spermatozoon (sperm cell), resulting in the first cell of the new individual. The nuclei of the egg cell and the sperm cell contain, respectively, the maternal and paternal genes, which carry the primary instructions that guide the formation and functioning of a living being. During fertilization, these two nuclei merge, combining the maternal and paternal instructions into a unique set of genes: the **genome** of a new individual. The cell and all its descendants will read that genome such as a recipe; the instructions written in each gene will be carried out to generate the new living being with all its specific characteristics (Figure 1.1a).

From then on, following the instructions in its genes, the first cell divides into two, those two in four, the four into eight, and so on, until reaching the trillions of cells that make up an adult person. At each cell division, the genome formed at fertilization is copied whole and passed on to each one of the daughter cells. In this way, with the exception of end-stage egg (ovum) and sperm cells, each one of our trillions of nucleated cells contains a complete copy of our genome.

By the fourth day of development, the human embryo is an amorphous conglomerate of 16 to 30 identical cells and is not yet implanted in the uterus (Figure 1.1a). By the fifth day, the embryo contains approximately 100 cells and is called the **blastocyst**. At this stage of development, some cells have now developed different characteristics,

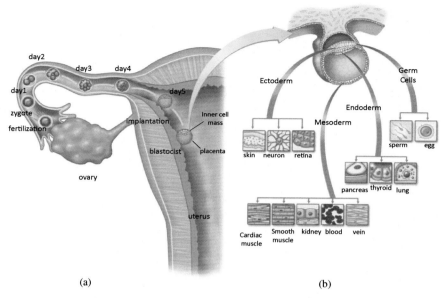

**Fig. 1.1.** Development of the human embryo. (a) With fertilization, the first cell (zygote) of a new individual is formed and begins to divide into identical cells. By the time it reaches the blastocyst stage, the embryo comprises two distinct cell populations: one that will give rise to the placenta and the other, the cells of the inner cell mass, that will give rise to the tissues of the new adult. (b) When the embryo cells implant in the uterus, they continue multiplying, and separate into four groups: ectoderm, mesoderm, endoderm and germ cells.

and divide basically into two groups: those that will form tissues outside the embryo, such as the placenta, and those that will form the embryo itself, which is referred to as the bud. In one or two days, the blastocyst expands and implants itself in the uterus, starting a complex process of cell divisions. From that moment progressively the cells in the embryo begin to take on specific forms and functions, in a process called **differentiation** (Figure 1.1b).

Differentiation means becoming different. Cell differentiation is the process by which otherwise identical cells become different from one another in structure and function. Remember that these few cells will give rise to all the structures in the new-born baby, from neurons and muscles through to skin, bone, liver and blood. If this process of specialization does not take place in a highly organized manner, it will not result in a normal functioning human being.

The first stage of differentiation divides the embryo into three large groups of cells called endoderm, ectoderm and mesoderm (Figure 1.1b), and a much smaller group comprising the germ cells. As the embryo develops, so the cells in each of these groups multiply and become progressively more differentiated and specialized into various tissues and organs.

The cells of the endoderm will produce the whole digestive system, liver, pancreas, kidneys and lungs. The cells of the ectoderm give rise to the skin and the nervous system, including the spinal cord and brain. Meanwhile, the cells of the mesoderm will form the connective tissues, cartilage, bones, muscles and the cardiovascular system, including the heart and blood cells. The germ cells, in the meanwhile, give rise to the gonads, either ovaries or testicles.

Initially identical cells are going to multiply and differentiate in an extremely organized fashion to produce a human being, made up of trillions of cells having more than 200 different cell and tissue types. But how do those initially identical cells acquire the specific characteristics of differentiated cells?

This process is governed by our genome, the instructions in the nucleus of each one of our cells. Thus, as the embryo develops, in some way that we do not yet fully understand, each cell begins to activate specific sets of genes, specific instructions in the genome: some will trigger particular genes to become blood cells; others, the genes of skin cells; some will turn on neuron genes, and so on (Figure 1.2). But how do the cells of the embryo know which genes to activate?

This is one of Biology's great mysteries and, to me, one of the most fascinating. Scientists have studied it by resorting to simpler organisms, such as *Drosophila melanogaster* (the fruit fly) or even the millimeter-long worm *Caenorhabditis elegans* (*C. elegans*), which contains exactly 1,031 cells in the adult male. Because that worm is so simple — and is more transparent — it is easier to study its development than in

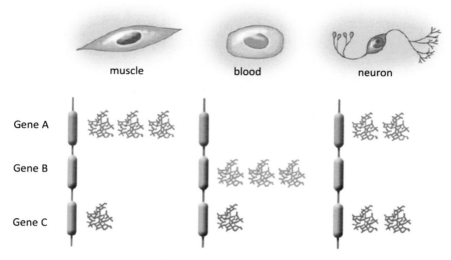

**Fig. 1.2.**   Identity of each cell type. Nearly all the cells in your body have a nucleus containing a complete set of all your genes, a complete genome. In each type of cell, however, only one subset of those genes is activated, giving the specific characteristics of each cell type.

higher organisms, and determine how it develops from the first cell to the adult individual.

[Watch a video of developing *C. elegans* at http://www.ib.usp.br/lance.usp/booksc/video1].

But what does the development of *C. elegans* or Drosophila have to do with human development?

Both started with fertilization, and continued with multiplication of initially identical cells, which at some point differentiated into specialized cells. It is as if, before trying to understand how to build a 100-floor building, the budding engineer starts by studying how to build a bungalow. The same basic laws of Physics that apply there will apply also to building the skyscraper — but it is simpler to learn them with a bungalow and then go on to more complex constructions.

In the same way, basic developmental mechanisms identified in *C. elegans* or in Drosophila are reproduced, although in more complex form, in the development of mammals such as mice and humans. This explains how knowledge obtained in these simple experimental animal models help us to understand the development of the human embryo.

To summarize, humans are made up of trillions of cells, each one usually containing a complete copy of our genome. Each type of cell will have activated a specific set of genes, thus becoming a specialized or differentiated cell in both form and function.

# 2 A Brief Introduction to Growing Cells Outside the Body, or Cell Culture

Chapter

With the exception of erythrocytes (red blood cells), each cell normally contains a complete genome in its nucleus. But are they able to live and multiply outside our body? Yes, but only if we provide them with the proper growth conditions, that is, nutrients, temperature and atmosphere that simulate the same environment surrounding them as inside the human body. Scientists had been researching how to do this since the 19th century, but it was not until the early 20th century that '**cell culture**' techniques emerged. By this process it is possible to induce cells to grow outside the organism, in culture vessels in the laboratory.

Cell cultures can be set up from small fragments or aliquots of skin, blood, liver, muscle and other tissues. The fragments are placed in flasks with a liquid containing essential nutrients for growth, the "**culture medium.**" These flasks are then kept at a temperature of 37°C in cell incubators containing a mixture of gases (importantly $CO_2$ and $O_2$). In this way, the cells from tissue fragments start to multiply, grow out from the initial fragments, and spread through the flask (Figure 2.1).

In the body, blood cells live in suspension and that is also how they are grown in cell culture in flasks. Cells of solid tissues, such as skin, muscle and liver adhere to the culture flask as they grow. Each cell type requires a specific culture medium, with varying levels of sugars, proteins and other nutrients and growth factors. In fact, one of the great challenges of cell culture is how to develop a culture medium suited to growing the desired types of cells.

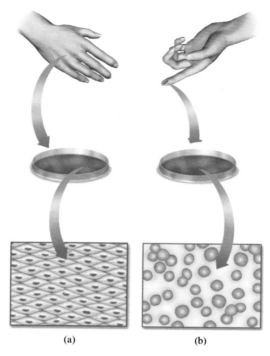

**Fig. 2.1.**   Culturing cells in the laboratory. Human cells can be multiplied in culture flasks and observed under a microscope: (a) skin cells; (b) blood cells.

[Watch a video of cells growing from a skin biopsy specimen at http://www.ib.usp.br/lance.usp/livroct/video2].

The ability to cultivate cells in the laboratory, or *in vitro,* has led to great advances in basic research and medicine. Cell cultures have been used to multiply viruses to produce vaccines, against poliomyelitis, for instance. From such cultured cells, it was possible to produce important proteins, such as insulin for treatment of diabetes or antibodies to fight specific types of cancer.

In principle this means that, if cells can be multiplied in culture, the problem of demand for tissues for transplants is resolved: all you have to do is take a little piece of the desired tissue and cultivate the cells to produce enough for a transplant. Right? Well, in theory, yes; but in practice, most types of cells, such as skin cells for instance, only multiply in culture for a limited number of generations and then stop growing. The same happens with all the other tissues that have been cultivated — cells that are already differentiated and specialized multiply up to a maximum of ~60 times, then stop dividing. Worse still, some cultures that do not stop growing have generally lost the ability to control their multiplication which, in an organism, often gives rise to tumor.

Is the limitation on cultured cell proliferation a technical problem (of discovering the right conditions for cell culture) or is it a problem of biology (that normal body cells actually cannot multiply forever)? Maybe both, but we do know that the great majority of cells in the body have limited ability to multiply. However, there is a special group of non-cancer cells that can multiply almost indefinitely both *in vivo* and *in vitro* and are responsible for the normal maintenance of all organs and tissues in the human body: these are known as stem cells.

# **3 Regeneration and Stem Cells**

All through life, the body is continually replacing cells of various tissues and organs. Take blood for example: its tissues are produced continually, to the point where we can donate blood regularly — we produce some 100 billion new blood cells every day! All our organs are constantly being regenerated: cells die and must be replaced by new cells; some, including skin, intestine and liver, much more often than others. Accordingly, there must be cells in the body that are not yet differentiated and specialized, and can take on the role of cell replacement in various differentiated tissues.

A stem cell (SC) is a cell with (1) a capacity for prolonged or unlimited multiplication — that is, it is able to divide many times over, generating cells identical to itself (Figure 3.1).

In addition, on receiving some outside stimulus, SC will (2) give rise to a more differentiated type of cell. Thus, a SC is a 'blank' from which differentiated cells can be generated, and at the same time maintain the stock of "blank" cells needed for future requirements.

The SCs responsible for replacing lost cells in specific organs are called **tissue-specific stem cells**. They have the ability to produce only cells of a specific tissue, be it skin, blood, heart or even brain. Those SCs can remain dormant, without dividing, for long periods, until the need arises to replace cells of that tissue — then the SCs multiply and differentiate to generate the specialized cells of that specific tissue. When the body is cut, for example, skin stem cells start to divide and produce the

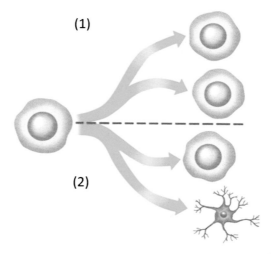

**Fig. 3.1.** What is a stem cell? To be considered a stem cell, the cell must be able to (1) divide into identical cells; and (2) give rise to different types of cell.

type of cells that make up skin: keratinocytes (cells with pigment, which give color to skin), cells of the epidermis and hair follicle cells. This is how a cut heals and regenerates the damaged tissue.

Since these specific SCs already exist in the body, they do not make us immortal, though they have the ability to regenerate different organs and tissues. The basic principle of regenerative medicine is to isolate and multiply specific SCs in the laboratory, so that, when transplanted into a patient, they can perform their intended function effectively and often under extreme situations, such as regenerating damaged heart tissue after a heart attack, repairing a broken spinal cord or diseased liver. However, since such cells only exist in relatively small quantities in the body, one of the great challenges facing regenerative medicine is to isolate and multiply the SCs from many different specific tissues, and to develop culture methods enabling the laboratory production of the large quantities of cells needed for therapy.

## Characterization of Stem Cells

Naturally occurring SCs can be divided into two large groups: **adult** and **embryonic** stem cells. Adult SCs form a large group including

SCs derived from any of an individual's tissues after birth. Accordingly, SCs from the blood of a new-born baby's umbilical cord or placenta, for example, form part of the "adult" group. Why divide stem cells into two groups? What are the differences between adult SCs and embryonic SCs?

# 4 Adult Stem Cells

## 4.1 Bone Marrow

### Hematopoietic Stem Cells — Our Blood Factory

The first SCs were discovered in the 1950s: these **hematopoietic SCs** reside inside the marrow of long bones. They give rise to all the cell types that are present in blood, such as immune cells, oxygen-transporting cells and the cells responsible for blood coagulation (Figure 4.1a). Hematopoietic SCs are thus classified as **multipotent SCs**, since they are capable of differentiating into various cell types, but are all derived from the same embryonic tissue, in this case, the mesoderm.

Thanks to hematopoietic SCs, blood has great capacity for regeneration, to the point that we are able to give blood regularly. When a person has a serious disease of the blood, such as a leukemia or anemia, a transplantation of bone marrow SCs can be performed. The transplantation procedure involves first destroying the host's own diseased hematopoietic SCs using chemotherapy and/or radiation therapy, and then infusing the bone marrow with SCs from a healthy donor. These will then start producing normal blood in the patient.

### Other Bone Marrow Cells — Mesenchymal Stem Cells

However, hematopoietic SCs are rare — in bone marrow, only one cell in every 1,000 is a SC capable of producing all blood cells. So, what is the function of other cells in bone marrow?

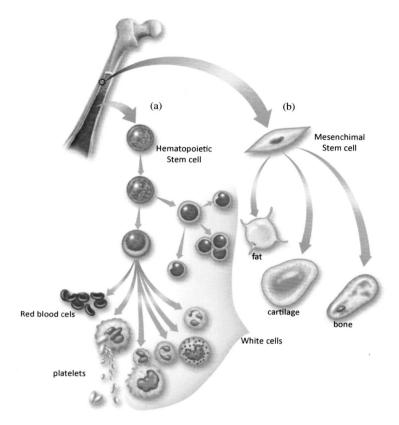

**Fig. 4.1.** Bone marrow stem cells. Inside the long bones are (a) the hematopoietic stem cells, which give rise to all the cells that make up blood; and (b) the mesenchymal stem cells, which give rise to bone, cartilage and fat.

Another type of SCs present in bone marrow are **mesenchymal SCs**, which serve as support for the hematopoietic SCs during blood formation and are able to differentiate into bone, cartilage and fat cells (Figure 4.1b). We now know that mesenchymal SCs are also present in other tissues, such as fat, dental pulp, placenta and umbilical cord vein. These mesenchymal SCs can give rise to bone, cartilage and fat, and so are also multipotent.

# Hematopoietic Stem Cells and Regeneration of Other Organs

In the late 1990s, starting with experiments in animal models, evidence began to emerge that, among the SCs in bone marrow, there might also be cells capable of regenerating organs such as the heart, liver and even the nervous system.

In 2001, researchers from one of these studies in the USA did the following experiment to test the versatility of bone marrow cells: they purified hematopoietic SCs from the bone marrow of a donor mouse, injected the purified cells into a recipient animal whose bone marrow had been destroyed, and then ascertained which organs the injected cells had ended up in (Figure 4.2).[1]*

However, how do you know which cells came from the donor's bone marrow and which were already present in the recipient animal? This is often done by using transgenic animals as donors whose genomes contain different genes, put there by human hand in a process referred to as gene transfer. In the 1980s various DNA "cut and paste" techniques were developed, known as recombinant DNA technology. With such methods, it became possible to isolate any gene from a species, whether bacteria, fly, plant or human, and insert it into the genome of another species (Figure 4.2a).

Why introduce a gene from one species into another? A gene is one instruction in the recipe for creating living things, from its genome. Each gene confers a specific characteristic to living things. For instance, the neo gene in bacteria is responsible for some bacteria's resistance to the antibiotic Neomycine; the F9 gene in humans produces the coagulation factor IX, which is essential for blood coagulation; and so on. Knowing how specific genes function and using transgenic techniques, a gene of interest can be inserted into the genome of another species, so

---

*The numbers in superscript refer to the references at the end of this manuscript.

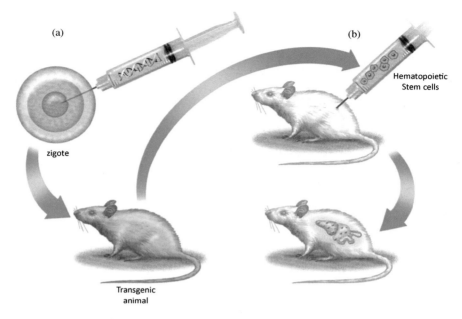

**Fig. 4.2.** Testing the versatility of hematopoietic stem cells. (a) Producing a genetically modified mouse: the GFP gene is injected into the mouse embryo, and is incorporated into its genome, causing the cells of the resulting animal to produce the green fluorescent protein (GFP); and (b) hematopoietic stem cells from the GFP mouse injected into a normal animal were found in several different organs.

that it acquires some desirable new characteristic. For example, plants can be supplied with a gene from a microbe that produces a protein toxic to caterpillars. In that way, the transgenic plants become toxic to caterpillars that attack them, and this solves a pest problem without resort to indiscriminate use of pesticides and wide spread contamination of the environment.

Returning to SCs, most studies of adult SCs used transgenic mice as bone marrow donors, especially animals containing a jellyfish gene responsible for producing a green fluorescent protein, known as the GFP gene. The presence of this protein in tissues is very easy to see and because of this the GFP gene is widely used to mark cells of interest. In this way, bone marrow cells from GFP mice fluoresce greenly, as do any other cells derived from the marked bone marrow. By using these transgenic animals as bone marrow cell donors, the cells and tissues derived from the bone marrow can then be detected by their fluorescence in the recipient animal (Figure 4.2b).

Some of the animals transplanted with fluorescent, transgenic bone marrow survived and as expected, their bone marrow was regenerated from the transplanted green fluorescent cells, which multiplied giving rise to blood cells in those animals. Nothing new so far: hematopoietic SCs from bone marrow normally go on to produce blood.

Now comes the really interesting part: some time later, the transplanted animals were sacrificed and the researchers detected cells inside them exhibiting fluorescence — that is, cells derived from the hematopoietic SCs injected into each animal. To everyone's amazement, although the fluorescent cells were largely in the recipient animals bone marrow, a small fraction of these cells was found also in various other organs, such as lungs, intestine and skin, indicating that hematopoietic SCs not only give rise to blood cells, but are also able to differentiate into cells of many other organs: hematopoietic SCs could thus be said to be **pluripotent** and capable of giving rise to any type of cell in the adult derived from the endoderm, ectoderm and mesoderm (Figure 4.2b).

This came as an enormous surprise to the researchers. Up to then, it had been thought that bone marrow only contained cells capable of producing blood, bone, cartilage and fat. The possibility that more versatile SCs might also exist in bone marrow enormously increased the therapeutic potential of the tissue, which until then had only been used to treat diseases of the blood. Accordingly, experiments were started in animal models to evaluate the therapeutic capacity of bone marrow SCs to treat various diseases. Bone marrow cells were injected into animals suffering from heart attacks, hepatitis, pulmonary emphysema, and even neurological diseases, in the endeavor to demonstrate the therapeutic effect of bone marrow cells on these various diseases.

This was a really interesting stage in research with adult SCs, with articles published in important journals showing that in animal models, bone marrow SCs had some ability to regenerate tissues in different organs. Bone marrow cells from a normal mouse, when injected into the heart muscle of another animal suffering a heart attack, turned into healthy heart muscle cells. When injected into the liver of a mouse with cirrhosis, those same cells now turned into hepatocytes — liver cells.

Note that in these studies, the bone marrow injected into the diseased animals was generally not purified. What was injected was a mixture of many bone marrow cells, including hematopoietic and mesenchymal SCs, and many other types of cells that also inhabit bone marrow, but about which little is known in detail. The cellular complexity of the injected bone marrow made it difficult to determine which types of cell were involved and how they achieved the observed clinical improvements. I will return to this again later.

## 4.2 Self-Renewal

Further studies in human subjects suggested that there are SCs in bone marrow with greater versatility than producing blood. One such study in 2002, examined men transplanted with hearts from women donors (Figure 4.3).[2] What was so special about these patients?

In terms of the human genome, men differ from women by having a Y chromosome — a chromosome that only exists in men. The patients in this study were men whose heart cells do not have the Y chromosome, because their transplanted hearts came from women. The presence of the Y chromosome in individual heart muscle cells was used to determine which heart cells came from the donor, and which from the recipient.

Some months after transplantation, biopsies were taken from these patients' heart muscles, and to the researchers' surprise, they found up to 12% of cells with the Y chromosome. Therefore, non-heart cells from the patient, containing the Y chromosome, had migrated to the transplanted heart and had been converted into new heart muscle.

One way of interpreting this result is to imagine that the body contains stocks of SCs responsible for repairing distressed organs or

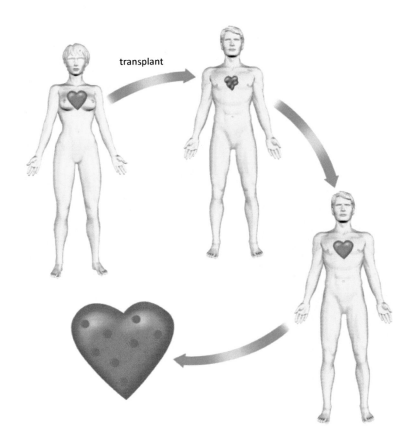

**Fig. 4.3.** Self-renewal of the heart. After a few months, men (XY) transplanted with hearts from women (XX) displayed some heart cells with the Y chromosome and hence derived from their own bodies.

tissues — and a transplanted heart is a distressed organ. On receiving a signal of suffering, these SCs are recruited for that organ and differentiate into heart muscle cells in an effort to regenerate new heart tissue. This means that there is a natural ongoing process of self-renewal in our body, which is responsible for maintaining our organs. However, this process is not robust enough to maintain the health of organs in extreme situations, such as a heart attack or spinal injury.

Now, knowing that self-renewal does exist in humans and animal models opens up new prospects to understand how organs can be

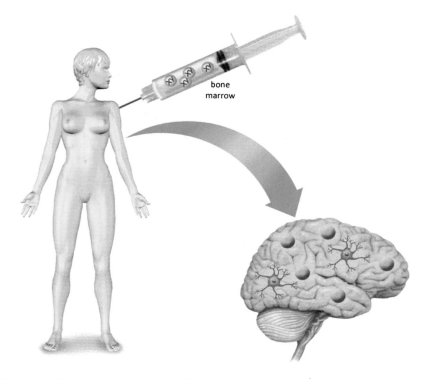

**Fig. 4.4.** Does bone marrow produce neurons? After some time, women (XX) transplanted with bone marrow from men (XY) display a few neurons with the Y chromosome.

repaired, to leverage already existing repair mechanisms and increase their efficiency, as occurs standardly in lizards and frogs, that can regenerate lost tails and limbs. Who knows, one day, instead of receiving transplants of SCs to regenerate a heart, we may be able to take a medicine that induces this self-renewal process by stimulating our own SCs to multiply, migrate to the heart and rebuild the heart muscle.

In 2003, another study with human subjects examined women with leukemia who had received bone marrow transplants from male donors (Figure 4.4).[3]

Once again the Y chromosome trick was used to distinguish donor cells from recipient cells: now women had bone marrow cells containing a Y chromosome. When the transplanted women died, analysis of their brains identified a small percentage (up to 0.07%) of neurons containing the Y chromosome — that is, derived from donor bone marrow. This study indicates the ability, although highly inefficient, of bone marrow cells to pass the blood brain barrier, to enter the brain and produce neurons, a phenomenon also observed in mice. If such ability could be enhanced, one day bone marrow SCs may come to be used to treat common neuro-degenerative diseases, such as Parkinson's and Alzheimer's.

## 4.3 Clinical Trials with Adult Stem Cells

All these findings have spurred research with a view to developing new therapies with bone marrow cells for common conditions such as heart attack, diabetes, liver cirrhosis and spinal cord injuries. In addition, as bone marrow transplants had already been performed for decades for blood disorders, it was known that those cells were at least safe and, accordingly, clinical trials involving bone marrow SCs were soon started in human subjects.

At http://www.clinicaltrials.gov (accessed on June 2015), the United States government describes clinical trials ongoing in the USA and elsewhere in the world to treat diseases in human beings, which comply with the respective countries ethical and health standards. A search in that database using 'stem cell' as search parameter returns more than a thousand ongoing clinical trials — the majority of which are varieties of bone marrow transplantation for diseases of the blood.

By refining the search, more than 300 trials were found to involve use of adult SCs to treat non blood-related diseases, including heart disorders, diabetes, spinal cord injury and multiple sclerosis. Some of the studies, such as those targeting epilepsy and stroke, are still in the initial phases, while others, directed at heart disease, are now at more advanced stages where it has been ascertained that the protocols are safe and the main goal is now to demonstrate their effectiveness in a large number of patients.

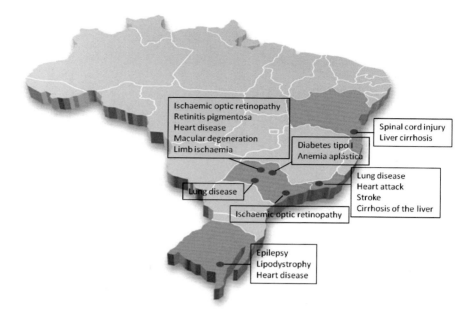

**Fig. 4.5.**   Clinical trials with stem cells in Brazil. Map of clinical trials with stem cells in Brazil (as recorded on clinicaltrials.gov up to June 2015).

In Brazil, the ministries of Health and of Science and Technology have funded a number of clinical trials with bone marrow SCs to treat various diseases (Figure 4.5). The largest has been conducted at more than 40 different research centers, involving 1200 patients with various kinds of heart disease and treated with SCs from their own bone marrow. This study, which began in 2005, included two groups of patients: one received injections of bone marrow cells into the heart muscle, while the other (placebo) group received an injection of saline solution containing no cells.

Neither patients nor doctors know which patients belonged to which group, in order not to influence the interpretation of the clinical measures taken before and after treatment. At the end of the study, the codes will be opened, and it will be possible to evaluate whether the group that received the cells improved significantly better than the group that received just the placebo — that is, it will be known whether this therapeutic strategy actually works.

********

## From basic research to clinical trials: the long road from conception to full implementation of a new therapy

**Basic research**: This is research performed to produce a basic understanding of human biology, with no intention of that knowledge being immediately applied to treating human disease. This kind of research can be done in a range of experimental models, ranging from bacteria, yeast, Drosophila, fish, mice to primates.

**Preclinical research**: From the information obtained from basic research, a new therapeutic strategy can be developed. The first tests of the safety and effectiveness of the new therapy will be performed on animal models, using the therapeutic agent (stem cells or a new substance) produced in a normal research laboratory.

If the results are promising, work will go on to a second phase of preclinical studies in which the therapeutic agent must now be produced in conditions appropriate for use in human beings, with stringent control of the reagents and manufacturing conditions, and once again tested in animal models.

If the results from using animal models are promising, an application to start trials in human subjects will be submitted to the regulatory agencies, presenting all the results from the preclinical studies.

- **Clinical trials**: Experiments with human subjects are categorized into Phase I, II, III or IV, according to the type of questions the study seeks to answer.
- Phase I clinical trials: Researchers test a new drug or treatment for the first time on a small group of people to evaluate its safety, determine safe variation in dosage and identify side effects.

- Phase II clinical trials: The drug or treatment is given to a larger group of people to assess its efficacy and evaluate its safety in greater depth.
- Phase III clinical trials: The drug or treatment is given to large groups of people to confirm its efficacy, monitor side effects, compare with other existing treatments and collect information enabling the drug or treatment to be used safely. Only after going through phase III will the new treatment be approved for wider use in patients.
- Phase IV clinical trials: These are studies conducted after the drug or treatment is on the market, to describe additional information, including risks, benefits and how best to use the drug or treatment.

********************

## 4.4 Other Sources of Adult Stem Cells

In addition to bone marrow, with its hematopoietic and mesenchymal SCs, other tissues also possess stem cells, and have been better characterized in recent years.

### Hematopoietic Stem Cells in Umbilical Cord Blood

At the end of the 1980s, it was discovered that blood from a newborn baby's umbilical cord and placenta is rich in SCs equivalent to the bone marrow hematopoietic SCs that give rise to all the cells that make up blood. These latter cells are formed in the fetus's liver and, close to birth, migrate from there to the bone marrow inside the long bones. Accordingly, hematopoietic SCs are still found in the blood circulating in the newborn, and even in the blood left in the placenta and in the umbilical cord when it is cut.

For this reason, rather than discarding this material, it can be collected and used, like bone marrow, to treat dozens of blood diseases, such as leukemia, lymphoma, anemia, as well as other human immune and hereditary diseases. Since this discovery, SC banks have been set up which now store thousands of cord blood samples for use in treating those diseases. Today, to find a compatible sample for transplantation, a patient can be referred to bone marrow donor banks and cord blood banks.

Nonetheless, in spite of all the effort to set up cord blood and bone marrow banks, the likelihood of finding a compatible donor or sample in such banks is generally not high. For this reason, some families opt to store their newborn's cord blood cells in private cord blood banks, either because they have some particular illness in the family or simply as a precaution. The difference between donating cord blood to a public bank and storing it in a private bank is that, while the sample donated to a public bank is anonymous and can be used both in research and to treat any patient, in a private bank only the family has access to the sample, which is perfectly compatible with the donor and offers an average of 25% likelihood of compatibility among siblings. On the other hand, while donating cord blood to a public bank is free, private banks charge for storage.

Is it worth paying for private storage? That discussion is subject to marked polarization of opinions: on the one hand, some private storage firms run sensationalist advertising, presenting cord blood stem cells as the cure for all ills and exploiting the moment when pregnant women, concerned with their new-born babies' health, are most vulnerable. On the other hand, perhaps as a reaction to this rather unethical marketing approach, some health professionals declare that there is no point to private storage, because the diseases treated with cord blood stem cells are so rare. Indeed, although bone marrow or cord blood cell transplants have been used to treat dozens of diseases, these are generally very uncommon diseases (Table 4.1). However, they are serious and often a compatible donor cannot be found in time. Whether or not such risk justifies private storage of cord blood is a very personal question, which should be discussed with a doctor bearing in mind, among other things, the existence of known diseases in the family and the extent to which the long term storage costs will burden a family's budget.

In any case, the identification of SCs in cord blood has revolutionized the fields of hematology and cancer treatment and opened up

**Table 4.1.   Diseases Treated by Transplants of Hematopoietic SCs from Bone Marrow or Umbilical Cord or Placental Blood**

Leukemias and lymphomas, including:

> Acute myeloid leukemia
> Acute lymphoblastic leukemia
> Chronic myeloid leukemia
> Acute lymphocytic leukemia
> Juvenile myelomonocytic leukemia
> Hodgkin's lymphoma
> Non-Hodgkin's lymphoma

Multiple myeloma and other plasma cell disorders
Severe aplastic anemia and other bone marrow failure conditions, including:

> Severe aplastic anemia
> Fanconi anemia
> Paroxysmal nocturnal hemoglobinuria
> Pure red cell aplasia
> Congenital / megakaryocytic thrombocytopenia

Other hereditary diseases of the immune system, including:

> Severe combined immunodeficiency
> Wiskott-Aldrich syndrome

Hemoglobinopathies, including:

> Beta thalassemia major
> Sickle-cell anemia

Hurler syndrome and other hereditary metabolic diseases, including:

> Hurler syndrome (MPS-IH)
> Adrenoleukodystrophy
> Metachromatic leukodystrophy

Myeloproliferative and myelodysplastic diseases, including:

> Refractory anemias (all types)
> Chronic myelomonocytic leukemia
> Idiopathic myeloid metaplasia (myelofibrosis)

Familial erythrophagocytic lymphohistiocytosis and other histiocytic diseases
Other malignant diseases

*Source*: National Marrow Donor Program, USA, 2012

new prospects of therapy for dozens of diseases. Furthermore, for the moment cord blood is the only alternative source of adult SCs now established for clinical use in diseases traditionally treated by bone marrow transplants. For treating more common diseases, such as heart attack and diabetes, cord blood continues in clinical trials along with bone marrow SCs.

One such study, conducted at Duke University, evaluated cord blood transplants to treat cerebral palsy caused by a lack of oxygen (hypoxia) in the brain. Only patients from 12 months to six years old whose cord blood has been previously stored could participate, because the transplant is performed with the child's own cord blood in what is called an autologous transplant. Experiments with animal models have shown that cord blood reduces the impact of brain damage caused by hypoxia. Although the mechanism behind this effect is still poorly understood, the researchers suggest that infusion of autologous cord blood cells in a patient could facilitate neural regeneration and thus improve the clinical condition of children with cerebral palsy. By 2015, studies had shown the procedure itself was safe, and an increasing number of patients were undergoing clinical trials to demonstrate the effectiveness of the procedure.

## Mesenchymal Stem Cells in Other Tissues

Lastly, mesenchymal SCs, equivalent to those of bone marrow, have been identified in a number of other tissues, including fat, placenta, dental pulp and umbilical cord vein. Similar to bone marrow SCs, all these mesenchymal SCs can differentiate into bone, cartilage and fat. Emerging evidence shows that there may be qualitative and quantitative differences between the different SCs in their differentiation ability that in some way make them different from bone marrow SCs.

Mesenchymal SCs derived from other tissues are also potential alternative sources of cells for therapy. Going back to the Clinical Trials database, in January 2012 there were a number of clinical trials using mesenchymal SCs isolated from fat for treating cardiac insufficiency, fistulas, spinal cord injuries, limb ischemia and so on. Mesenchymal SCs from umbilical cord vein are being tried in humans for treating liver cirrhosis, and others derived from placenta, for treating pulmonary fibrosis and ischemic cerebrovascular accident (stroke). These trials are still in the early phases, and it will be important to monitor their progress.

## Brain Stem Cells

One special class of adult SCs are **neural SCs**, which are derived from the brain. For many years it was believed that adult brain cells did not regenerate and that there was no cell replacement in the human brain. However, in the 1990s a small population of brain cells with SC properties was identified. These give rise to two groups of cells found in the nervous system: **neurons**, the functional units of the nervous system, which transmit information to other neurons or to other cells; and **glia** or glial cells, which support neurons.

For this reason, neural SCs have great potential for treating various types of neurodegenerative diseases, such as Parkinson's disease. However, as they exist in only small quantities in the organism, to be used in therapy they must be isolated from the brain and multiplied in the laboratory without losing their identity as neural SCs — that is, they must continue to be able to divide and differentiate into the various types of cells of the nervous system.

At least three companies, two in the USA and another in the United Kingdom, have established lineages of neural SCs from human fetal brain. They have isolated a small population of these cells from the brain and developed a method to multiply them in culture to generate billions of neural SCs, which theoretically could be used to treat several patients.

Wait a minute, from fetuses? Yes, abortion is legal in those countries and so there is access to this kind of material, provided appropriate consent is given by the surveillance agencies. As the brain in the fetus is still being formed, it contains large numbers of actively growing neural SCs which, as they are "younger," have the capacity to multiply efficiently in laboratory cell cultures.

These neural SCs are currently being tested in patients with spinal cord injury, amyotrophic lateral sclerosis (ALS, or Lou Gehrig's disease), stroke, macular degeneration and two genetic neurodegenerative diseases. These studies, still in phases I and II in 2015, are designed to ascertain the safety of using the cells, and whether they in fact are able to regenerate nerve tissue in the diseases concerned.

## Heart Stem Cells

Just as neural SCs, there are other types of tissue-specific SCs for other organs including liver, heart and intestine that are responsible for renewing the respective organs. Other types of tissue-specific SCs far less well known than neural SCs have now been identified, the same strategy is planned of isolating and growing them for use in regeneration of damaged organs.

In particular, major advances have been achieved with **cardiac SCs** isolated from small biopsies of heart muscle. Initially identified in rats, cardiac SCs are multipotent cells, which differentiate into the various different types of cells that make up the heart, forming heart muscle and blood vessels.[4] These same cells isolated from human heart tissue are responsible for heart maintenance and the small amount of self-renewal observed following heart disease. In some animal models, human heart SCs are able to help regenerate the heart after a heart attack.

Once those cells from heart biopsies have been isolated and multiplied in the laboratory, and shown to be effective in animal models, the trials in human subjects begin. In two such studies in the USA, one million SCs produced from heart biopsies of patients who had suffered heart attacks were infused into the coronary artery on the damaged side of the heart.[5,6] After 6 months of treatment, the patients treated with their own heart SCs showed significant improvement in heart tissue and contractile capacity as compared with the untreated patients. In addition, one of the studies reported that heart function improved substantially in patients treated with non-autologous heart SCs. These experiments in human subjects indicate that therapy with heart SCs is safe and leads to regeneration of the heart. Now the strategy is to be tested on a larger number of patients to confirm efficacy in improving their clinical condition.

## Germline Stem Cells

Other organs of great interest to medicine are the gonads, ovaries and testicles, producers of ova and spermatozoa, respectively. Production of these cells can be lost or impaired as a result of various situations, such as advancing age, particularly in the case of women, or exposure to chemotherapy or radiotherapy as cancer treatment in either sex.

To overcome these situations, some assisted reproduction techniques have been developed, particularly *in vitro* fertilization. However, for this to be successful it requires both eggs and spermatozoa — can cells that are so specialized be produced from SCs?

Germ cells are unique, because it is from them that new individuals are produced. As seen above, each germ cell has only half the individual's genes. So, when a spermatozoon fertilizes an egg cell these two cells merge, the two halves join together to form a new genome, a complete new recipe for a unique individual in the world.

Remember when the embryo starts to develop, when it is no more than a conglomerate of identical cells? At the start of the process of differentiation, the cells divide into endoderm, ectoderm and mesoderm, a small group of cells is held in reserve that go on to form the gonads (ovaries or testicles), and in puberty those organs begin to produce mature germ cells.

Until recently, it was believed that egg cells were produced only until birth. Women were born with a fixed number of egg cells, which matured from puberty through the fertile phase — on average, one egg cell was readied for fertilization per month, and a woman would ovulate on average 400 times. However, in 2004, a group at Harvard University, in the USA, identified a population of cells (**germline stem cells**) that divided and formed new egg cells in the ovaries of adult mice.[7]

That discovery stimulated research into human reproduction, opening up new prospects for treating female infertility. If these cells also exist in humans, then can they be multiplied in the laboratory to produce egg cells in large quantities? Imagine the impact that this could have on prolonging female fertility!

All revolutionary discoveries spark controversy, and the idea that there exists such a thing as female germ SCs met resistance from many research groups. However, in 2009 a group at Jiao Tong Shanghai University, in China, managed to isolate germ SCs from the ovaries of newborn and adult mice, and to multiply those cells in culture.[8] In order to demonstrate that the cultivated cells were in fact germ SCs, the researchers inserted the GFP gene into them, which produces fluorescent jellyfish protein, and transplanted the cells into the ovaries of sterile mice. In these animals, the transgenic germ SCs gave rise to egg cells, and the hitherto sterile mice had offspring whose genome contained the GFP gene — proof that these baby mice were generated from egg cells derived from the transplanted germ SCs.

Finally, that Harvard group which, in 2004, had identified female germ SCs in mice and then suffered years of skepticism from some colleagues, managed to isolate a small cell population from human ovaries, which when multiplied in culture were capable of giving rise to egg cells.[9] In addition, when these cells were transplanted into the ovaries of immune deficient mice, the researchers also observed human egg cells form in those animals.

Baptized **oogonial (egg-making) SCs**, they give rise to only one type of differentiated cell (egg cells), and are thus classified as **unipotent SCs**. Of course, a lot more research is needed to demonstrate that human egg cells produced from these cells are functional, can be fertilized and give rise to a normal baby. But this work may be an important step towards increasing female fertility. If the oogonial SC results described above are confirmed, we may be able to generate an unlimited number of egg cells from biopsies of the ovary, which would solve the main problem facing women treated at assisted reproduction clinics. In addition, the ability to produce egg cells from oogonial SCs in the laboratory might make it easier to identify drugs and hormones that stimulate these cells in the organism to produce egg cells for longer, thus slowing a woman's biological clock.

In contrast with the limited natural production of gametes in females, males produce billions of spermatozoa per month from puberty onwards for the rest of their lives! This enormous output is maintained by a small population of stem cells in the testicles: these cells divide into other cells identical to themselves (thus maintaining the blank from which to produce more spermatozoa) and also into cells that give rise to the mature spermatozoa. These stem cells are called **spermatogonial SCs**, and as they give rise to only one type of differentiated cell (spermatozoa) they too are classified as unipotent SCs.

Spermatogonial SCs were isolated from mouse testicles in 1994, and their ability to generate functional spermatozoa was demonstrated by transplanting them into the testicles of sterile animals, similar to what was done with female germ SCs.[10] In 2010, spermatogonial SCs were isolated from human testicles and multiplied in culture.[11] Then, in 2014, another group was able to isolate spermatogonial SCs from testis of patients that did not produce sperms.[12] Moreover, they developed methods to multiply those SCs in culture and to differentiate them into spermatozoa. If it is proven clinically that the spermatozoa derived from human SCs can generate normal human embryos, they may provide an option for treating some forms of male fertility.

This approach can also be valuable for male cancer patients. It is true that men who are about to undergo chemotherapy, a treatment that may damage and even destroy their germ cells, already have the option of freezing their spermatozoa beforehand. However that strategy will not work in cases of cancer in boys who have not yet entered puberty, since their testicles do not yet produce spermatozoa. However, if their spermatogonial SCs could be extracted, multiplied and frozen, then those cells could later be reinjected into their testicles and normal production of spermatozoa induced in those patients.

## 4.5 Cancer Stem Cells

One of the most terrifying things a doctor can announce to a patient is: "Your cancer has come back...." Cancer is disorderly cell growth, cells multiplying when they should not and producing a tumor. Why, after surgical removal, radiotherapy and chemotherapy, do tumors reappear out of nowhere? The most likely explanation is that some cancer cells have survived all those attacks and are able to give rise to another tumor — these resistant cells "regenerate" that tumor. Now doesn't that remind you of the definition of stem cells? Cells with the ability to proliferate for long periods and able to give rise to other types of more differentiated cells... Except that, in this case, the stem cell in question is not giving rise to a normal tissue, but rather to cancer — it is a cancer stem cell.

Cancer SCs were first identified in leukemia's (blood cancers) in 1997.[13] By transplanting patient leukemia cells into immune-deficient mice, researchers demonstrated that not all the diseased cells gave rise to a leukemia in the animals. Only a small fraction of those cells were able to recreate the disease in the mouse — that ability to recreate the same cancer is what defines a cancer SC. Since then, cancer SCs have been identified in a variety of tumors including breast[14] and brain.[15] If such cancer SCs actually exist, and if they can be killed, that would be cutting the tumor off at the root and it would not return.

However, although the concept of cancer SCs makes sense, and there is evidence of such cells in other types of tumor, controversies still exist in the area. Some researchers for instance, found that only

0.0001% of cells in a tumor are cancer SCs, while other groups found up to 25% of cancer SCs in the same type of tumor. Who is right? Cells turn cancerous when their genes undergo a series of mutations and the cells begin to divide and grow in an uncontrolled fashion. So, do cancer SCs arise from any type of cell or from normal SCs that undergo mutations?

Cancer is a very heterogeneous group of diseases: cancers in different organs have different characteristics, and even in the same group (breast tumors, for example) there is enormous variation, ranging from benign tumors through to the most aggressive kinds, which form metastases. Can it be that cancer SCs are the origin of all types of cancer?

Over the course of history, various other promising theories for cancer have emerged; not all of them have resulted in more efficient therapies. Whatever their clinical potential, the concept of cancer SCs has made researchers think about cancer in innovative ways — and it is always useful to look at a problem from different angles and help knowledge advance.

## 4.6 Mechanisms of Action of Adult Stem Cells

One important issue in therapies using adult SCs is to understand exactly the mechanisms by which they produce any beneficial effect on pathologies. The therapeutic effects of tissue-specific SCs are well understood: neural SCs give rise to the cells of the nervous system; cardiac SCs differentiate into the different heart cell types; and hematopoietic SCs from bone marrow and umbilical cord blood give rise to all blood cells. When transplanted, all these SCs renew the respective damaged tissue. But what about the other cells present in bone marrow, such as mesenchymal SCs, which become bone, fat and cartilage? Are they really differentiating into other types of cells and thus renewing tissues such as heart, liver and brain?

Although early studies suggested there were bone marrow SCs capable of producing other tissues besides blood, other studies questioning

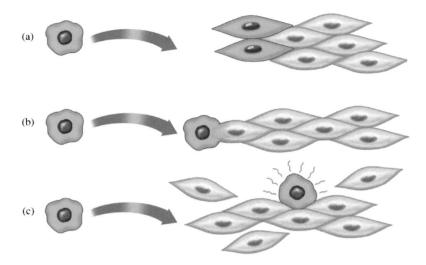

**Fig. 4.6.** Mechanism of action of bone marrow stem cells in organ renewal. Many studies show that, instead of (a) turning into cells of the various different organs, bone marrow SCs (b) merge with the cells of the respective organs or (c) secrete factors that foster the organs' self-renewal.

that model soon appeared. Some indicated that when these SCs were injected, rather than turning into cells of the diseased organ, they merged with the cells of the tissue and, in that way, appeared to have turned into cells of that tissue (Figure 4.6). Other studies showed that, in fact, the transplanted SCs secrete growth factors, signal proteins that recruit the patients' own cells to effect self-renewal of the organ — which, moreover, is the most widely accepted hypothesis.

Also, in most of the studies using bone marrow SCs for organ renewal, a mixed population of cells is aspirated from inside the long bones. Now, which cells in that heterogeneous population are performing some therapeutic effect? The fact is that, at present, few people believe that mesenchymal SCs are able to give rise to all the tissues that were originally imagined. However, they do seem to have some therapeutic effect on some diseases, but how?

Clinical research with mesenchymal SCs is at an empirical stage: the cells are infused into patients to see if they show any improvement, but without knowing exactly which bone marrow cells produce which therapeutic effects. Based on the current view that mesenchymal SCs are unable to differentiate into anything except bone, cartilage and fat — which, ironically, was the initial view — there is no longer any strong scientific justification for clinical trials using such cells to treat, for instance, neurological diseases. All the same, such trials in human subjects continue — and they should continue to do so, because even though there is currently no logical explanation, mesenchymal SCs do seem to occasionally have therapeutic effects on some diseases, and that must be explored in greater depth.

Lack of basic knowledge does not necessarily prevent clinical trials from advancing — after all, simple drugs, such as aspirin, were used for many years before anyone knew exactly what mechanisms they acted by. However, there has to be simultaneous investment in basic research to understand exactly which bone marrow cells foster the improvements observed in some clinical trials, and also what mechanisms enable them to produce therapeutic effect on the various diseases they are being tested on. That knowledge will make it possible to boost the effectiveness of such treatments and to develop new cell therapy strategies for other diseases.

A study published in 2011 on the therapeutic mechanism of bone marrow SCs in myocardial infarction (heart attack) illustrated very clearly how important basic research is to cell therapy.[16] In the early 2000s, some studies had concluded that, when bone marrow SCs were injected into mice that had suffered heart attack, they turned into heart muscle cells, thus renewing the organ. When a group at Harvard University examined this phenomenon more closely they discovered two very important things about that kind of therapy: firstly, rather than turning into heart muscle, the bone marrow SCs fostered self-renewal by the muscle. That is, they somehow induced the heart SCs already present in the organism to multiply and differentiate into heart muscle cells. Secondly, it is in fact a subgroup of bone marrow SCs that has this effect of inducing self-renewal — and these cells, which produce a protein called c-kit, are only a very small fraction of all bone marrow cells.

If these findings are confirmed by others — and in science this is absolutely essential if new knowledge is to become established truth — they will be crucial to developing new SC-based heart attack therapies. Once it is established exactly which group of bone marrow cells is producing the therapeutic effect, it will be possible to redesign clinical trials in human subjects for cell therapy specifically using those cells.

Meanwhile, knowing that the injected cells, rather than turning into heart muscle themselves, send out signals to the patient's own cells (in this case, still a mouse) to renew the heart, it may be possible ultimately to do without the inducing SCs altogether and develop drugs to carry out the signaling. That, of course, will entail identifying what the self-renewal signals are — but the important thing is elucidating the mechanism by which administration of bone marrow SCs fosters improvement to the heart muscle and making it possible to work more intelligently to develop a really effective therapeutic strategy to treat heart attack.

## 4.7 Mesenchymal Stem Cells and Control of the Immune System

One particular class of diseases that are potentially treatable with SCs are autoimmune diseases. Remember that the immune system consists of blood cells that normally protect against agents from outside the body.

As the name suggests, in autoimmune diseases, the patient's immune system fails to recognize certain organs or tissues as its own, and attacks them as if they were enemies. One example is Type I diabetes: the patients have normal pancreas cells, which produce insulin; however, those cells are destroyed by their immune system, leaving the individual deficient in insulin and prone to develop diabetes.

Auto-immune attack of tissues is believed to be the mechanism behind multiple sclerosis, erythematosus lupus, rheumatoid arthritis and several other auto-immune diseases.

Even though not knowing what causes an auto-immune system defect, some scientists raised the hypothesis that, in the same way as rebooting a stalled computer, the immune system may be restored to normally functioning if it could be 'rebooted'.

The immune system is produced from the hematopoietic SCs found mainly in bone marrow or in cord blood. So, in order to 'reboot' the system, a small quantity of the patient's bone marrow SCs is withdrawn and stored. Meanwhile, the patient's existing (defective) immune system is destroyed with chemotherapy. To renew it, the patient's own stored bone marrow is then re-injected. It will produce new immune system cells, and the hope is that these will recognize the patient's pancreas as their own and not attack it.

For example, in the case of diabetes it is important to know exactly the stage of disease progression when the disease was diagnosed to be able to predict whether or not this treatment will be successful. This rebooted SC transplantation does not form new insulin-producing cells; it just interrupts the immune system's attack on the pancreas cells. Therefore, if the patient has had the disease a long time, their immune system may have already destroyed a large number of those cells, and stopping the attack will not be of much use since not enough cells will remain to produce the necessary quantity of insulin.

This strategy of "rebooting" the immune system works in some cases, but not in others — and it is not yet known why patients differ in this respect. In all cases, as the treatment involves destroying the immune system, it leaves the patient vulnerable to infections for several days until their defenses are renewed. This represents a major mortality risk and in fact, in early clinical trials, some patients did not tolerate the treatment, and died from infection.

With this and other clinical studies of bone marrow SCs, it was discovered that mesenchymal SCs — those that give rise to bone, fat and cartilage — act on the cells of the immune system to make them less active — that is, the mesenchymal bone marrow SCs seem to have an immunosuppressive effect.

As a result, a new therapeutic strategy for autoimmune diseases is being tested with mesenchymal SCs. Without destroying the patients' immune system and thus exposing them to great risk of death, mesenchymal SCs purified from bone marrow are multiplied in the laboratory and injected into the patient's bloodstream, which in turn, leads to a reduction in the activity of their immune system, thus the immune system does not attack the host body.

The immunosuppressive effect of the mesenchymal SCs purified from bone marrow, cord blood or even from fat is also being explored to treat other autoimmune diseases, such as multiple sclerosis and lupus. Additionally, as these cells are not recognized as being foreign and attacked by the immune system of the host, mesenchymal SCs from any donor can potentially be used.

It is still not known how these cells escape being attacked and control the patient's immune system, nor whether they will in fact manage to do this in the long term. Nonetheless, this is another developing area for SC therapies that does not involve renewal of organs or tissues, but exploits the mesenchymal SCs potential for regulating immune system activity.

## 4.8 Adult Stem Cells: Conclusions

Since the late 1990s, knowledge of adult SCs has evolved in very interesting ways. From being cells with a limited ability to differentiate (bone marrow only makes blood; mesenchymal cells only make bone, cartilage and fat), now — with their apparent ability to differentiate into various other types of cell — they have become the stars of regenerative medicine.

After a little more research, it soon became clear that the initial results could have other explanations, and today few people still believe that the majority of adult SCs are all versatile.

Over thirteen years of research, developments have come full circle and returned to the initial model in which the various types of adult SC are regarded as having limited ability to differentiate. Studies are in progress to identify which subgroup of bone marrow cells actually has some therapeutic potential, and in discovering the real mechanisms by which they apparently foster improvement in some conditions, such as autoimmune diseases.

It has identified various tissue-specific SCs, such as neural, cardiac and germ SCs, which really are able to regenerate the respective organs. In addition, the therapeutic capacity of some types of adult SCs has now been tried in human subjects.

The first clinical results with bone marrow SCs may not have been as promising as anticipated in the early 2000s, but a lot has certainly been learned over the years. That knowledge will provide the basis on which research with adult SCs can continue to explore all their therapeutic capacity.

# **5** Embryonic Stem Cells

The second major group of SCs, known as **embryonic stem cells**, differ significantly from adult SCs. As their name suggests, embryonic SCs are derived from embryos. Identified in the early 1980s in mice, these cells are extracted from embryos that have developed for three days — then known as blastocysts (Figure 5.1).[17] Remember that at this stage of its development the embryo comprises approximately 100 cells, which divide into two types: those that will give rise to the placenta and those that will give rise to all adult tissues, the cells of the embryonic 'bud' or inner cell mass (ICM). These latter, in a continuous process of divisions and specializations, will develop into muscle, neurons, liver, and cells of all our organs and tissues. However, at this point of very early development, it is not yet decided what sort of tissue or organ they are going to turn into and, as a result, they are enormously versatile or **pluripotent**.

The inner cell mass can be removed from the blastocyst and multiplied in the laboratory in special culture conditions so that the cells retain their extraordinary ability to turn into any cell type (Figure 5.1). In this way, these wildcard cells can be produced in large quantities in the laboratory and offer a potentially unlimited source of cells for transplants. Note that, while adult SCs can give rise to only certain tissue types, embryonic SCs are able to give rise to all cell types in the human body.

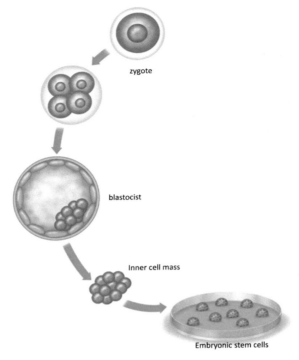

zygote

blastocist

Inner cell mass

Embryonic stem cells

**Fig. 5.1.**   Embryonic stem cells. The cells of the embryonic bud are removed from the blastocyst and placed in a culture flask, where they multiply and give rise to embryonic stem cells.

How can it be demonstrated that these millions of embryonic SCs remain pluripotent after removal from the embryo and expansion under artificial *in vitro* conditions in the laboratory?

The most convincing way is to reintroduce such embryonic SCs into a natural blastocyst and then examine the resulting animal to see whether some tissues were produced from the introduced embryonic SCs (Figure 5.2a). That experiment is called **production of chimeras** (that is, animals made up from two cell types from different genetic origins — in this case, cells from the original blastocyst and embryonic SCs injected into the embryo). If the introduced embryonic SCs really are pluripotent, they are incorporated into the tissues of the embryo

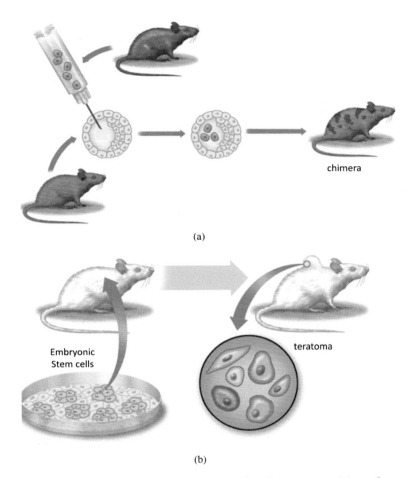

(a)

(b)

**Fig. 5.2.** Demonstration of the pluripotency of embryonic SCs. (a) Production of chimeras: embryonic stem cells derived from a black-furred animal are injected into the embryo of a brown-furred animal. If the embryonic stem cells are pluripotent, they will give rise to the various tissues of the resulting chimera. (b) Formation of tumors: embryonic stem cells injected into immune-deficient mice form teratomas, which are tumors made up of various different tissues.

and, during development, will differentiate into many types of tissue in the animal, including germ cells.

Another way of demonstrating that the embryonic SCs are pluripotent is to inject these cells into immunodeficient mice (Figure 5.2b). In the mouse organism, embryonic SCs receive stimuli to differentiate into many specific cell types. However, because they are 'wildcards', they are able to respond to the many different stimuli, which launches them into a chaotic process of random specialization giving rise to a benign tumor called a teratoma, and containing various different types of tissue (muscle, intestine, neurons, blood vessels, etc.). This implies that since the embryonic SCs injected into the animal had the ability to turn into so many types of tissue, they really were pluripotent. This strategy is used in animal models to demonstrate the pluripotency of embryonic SCs of species (such as humans) where ethical considerations preclude the production of chimeras.

However, if embryonic SCs are to be used as a source of tissues for transplants, one has to be certain that they will form organized normal tissues and not disorganized tumors in the recipient. Accordingly, before they can be used for transplantation, it must be experimentally determined how to differentiate them into the required cell and tissue types.

Now, the great natural aptitude of embryonic SCs is to differentiate — to continue multiplying without differentiating they have to be kept in very special culture media in the laboratory. When we want to turn those pluripotent cells into specialized cells, all that has to be done is to place them in the same medium as used to culture other types of cells (skin cells, for instance).

Without the restraints of the culture medium mentioned above, the embryonic SCs embark on a process of differentiation in the laboratory similar to normal embryo development. After two weeks, those pluripotent SCs will differentiate into neurons, skin (fibroblasts), muscle and cardiac muscle cells, which form clusters contracting rhythmically in the culture flask similar to a sheet of heart muscle (Figure 5.3).

However, for therapeutic use, more homogeneous cell populations must be produced to suit the specific diseases to be treated or organ to be regenerated. That is the great challenge of therapies using embryonic stem-cells: how to tame them in culture so that they turn into only the desired cell type.

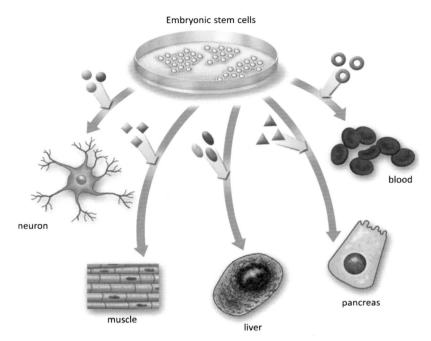

**Fig. 5.3.** Steering the differentiation of embryonic stem cells. By culturing embryonic stem cells with different compounds they can be induced to differentiate into specific cell types.

[Watch a video of embryonic stem cells transformed into heart muscle cells contracting synchronously in the laboratory. Available at: http://www.ib.usp.br/lance.usp/bookct/video3].

Accordingly, over the past 30 years, embryonic stem cell research has developed methods to culture and multiply SCs and turn them into specific cells of bone marrow, heart muscle, neurons and so on. Many articles describe ways of producing homogeneous populations of differentiated cells derived from embryonic SCs. For instance, addition of retinoic acid to the culture medium at defined concentration causes embryonic SCs to turn into neurons and addition of activin A results in the formation of cardiomyocytes (heart muscle cells).

How do we know how to steer the differentiation of embryonic SCs into a specific cell type?

Scientists have had to rely on what they have learned about embryo development, basically from model organisms. During the embryo development, all these types of cell differentiation take place in an orderly fashion. The endeavor then is to reproduce, in culture flasks, what takes place in the region of the embryo that will become, for instance, the brain, and so turn undifferentiated cells into neurons.

One example is how the methodology to turn embryonic SCs into insulin-producing pancreatic cells has developed. Pancreas cells were known to originate from the group of differentiated cells called the endoderm (Figure 1b). Studies in mice had also identified a series of proteins involved in the process by which some of the endoderm cells were converted into pancreatic cells. A research group from a United States company recreated these conditions in the culture media of embryonic SCs by adding the various proteins sequentially, over the course of 20 days, to guide the differentiation of pluripotent cells into insulin-producing cells.[18]

Of course, scientists are not able to reproduce in cell cultures everything that takes place in the organism, and these studies involve a considerable measure of trial and error as well. The important thing here, though, is to stress the significance of this interaction between basic knowledge of developmental biology and research applied to cell therapy.

Now, do these cells derived from embryonic SCs have any therapeutic effect in living creatures?

Yes, they do. A number of studies report that, when differentiated embryonic SCs are transplanted into animal models, they can relieve the symptoms of various diseases, from diabetes and Parkinsonism to paralysis caused by spinal cord injury. Contrary to mesenchymal SCs, where the mechanism of their therapeutic action is unknown, embryonic SCs are known to act by integrating and multiplying in the diseased organ or tissue and regenerating it.

## 5.1 The Embryonic Stem Cell Controversy

In 1998 the first lineages of embryonic SCs derived from human embryos were produced at the University of Wisconsin, in the USA.[19] Based on methods used to culture mouse embryonic SCs, these researchers managed to develop a suitable culture medium for human embryonic SCs extracted from human blastocysts. Now, where did those embryos come from?!

From *in vitro* fertilization clinics. When a couple resorts to assisted reproduction methods (ART), particularly *in vitro* fertilization (IVF), the woman receives hormone injections to produce a large number of mature follicles at once. Oocytes are removed from the follicles by suction in the laboratory and brought into contact with spermatozoa to facilitate fertilization. The resulting embryos are cultured up to the 4–8 cell stage reached after ~3 days, or at most until they are blastocysts at ~6 days. Then, one to three of these embryos are transferred to the uterus.

Three? Is the couple going to have triplets? Probably not, but it is possible. The average rate of implantation of each embryo implanted into the uterus is around 30%; so it is most likely that, of the three embryos transferred, only one will implant and develop into a baby.

As the process of *in vitro* fertilization is not 100% efficient, usually it begins with 6 to 10 follicles for both ovaries per IVF cycle (this will depend on how many follicles develop after hormone stimulation). Of these, up to 70% of the isolated oocytes will be successfully fertilized, generating 4 to 7 zygotes, which will be cultured until they reach the 4 to 8 cell stage. Another 20% of the embryos are lost at this stage, leaving 3 to 6 embryos ready to be transferred to the uterus.

So, often more than three embryos are produced that can be transferred. However, the couple may not want to run the risk of a multiple pregnancy and choose to transfer two or even just one embryo.

Then what is done with the other, leftover embryos? The surplus embryos may be discarded (depending on the country's laws — in Brazil, it is illegal to discard human embryos), frozen for the couple's future use, donated to another couple who cannot produce embryos of their own or even donated for research.

It was with such surplus embryos from IVF that the Wisconsin researchers worked, culturing them into blastocysts, removing the embryonic buds and adapting those cells to laboratory culture conditions. Considering that mouse embryonic SCs first appeared in 1981, it took considerable time before scientists managed to do the same thing in humans. Of course, it is a lot easier to produce large quantities of mouse embryos than human embryos. Not only that, but culture conditions for human and mouse embryonic SCs are not identical, and it took some time to discover how to culture embryonic SCs from humans.

The important thing is that everything the scientific community learnt from mouse embryonic SCs about controlling differentiation, it has applied to human embryonic SCs. Scientists now know how to induce the transformation of human embryonic SCs into various cell types including neurons, blood cells, liver cells and even heart muscle cells that contract in the culture flask. There are also numerous articles describing how, just as with mouse cells, human cells have important therapeutic effects in animal models of various diseases.

However, with the emergence of embryonic SCs derived from human embryos, there also emerged a new obstacle to be surmounted for those cells to be put to therapeutic use: the controversy over the destruction of human embryos as removing cells from the as bud of a blastocyst to obtain embryonic SCs does indeed destroy that human embryo.

Careful though: when you say "human embryos", people tend to imagine a fetus in human form, heart beating. Here, we are talking about much younger embryos only 5 days old, comprising approximately 120 cells, those surplus embryos from *in vitro* fertilization, which have not yet even been implanted in a uterus (Figure 5.4). Even so, to some people, destroying those embryos is equivalent to killing people, and that is an unacceptable price to pay for embryonic SCs. This issue has been the

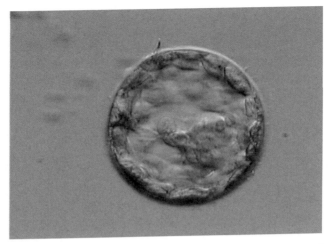

**Fig. 5.4.** Human embryo from where embryonic stem cells are removed. Photo of a blastocyst (a 5 day old human embryo) produced by *in vitro* fertilization.

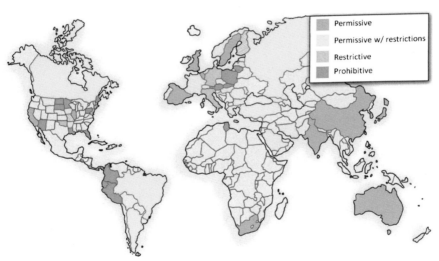

**Fig. 5.5.** The restrictiveness of countries' laws on use of human embryos for research. More details on each country's laws at <http://www.hinxtongroup.org/wp.html>.

subject of intense debate across the world, and each country has decided how to use embryonic SCs according to its own laws, culture, religion and history (Figure 5.5).

# The Controversy in Brazil

In Brazil, the controversy over using human embryos was settled in the 2005 Biosafety Law, which allows embryos to be used for research and therapy if they have been produced by *in vitro* fertilization, are donated with the parents' consent and have been frozen for at least three years, so that the parents had good time to think before making the donation.

## *Bio-safety Law — No. 11,105, 24 March 2005*

Art. 5: Embryonic stem cells obtained from human embryos produced by *in vitro* fertilization and not used in the respective procedure may be used for research and therapy purposes, under the following conditions: I — they are unviable embryos; or II — they are embryos that have been frozen for three years or more at the date of publication of this law or which, if already frozen at the date of publication of this law, after completing three years counted from the date on which they were frozen.

§ 1: In any case, the genitors' consent is required.

§ 2: Research institutions and health services that conduct research or therapy with human embryonic stem cells shall submit their projects for consideration and approval by the respective research ethics committees.

Shortly after it was sanctioned, however, an attorney-general filed an action challenging the constitutionality of the 2005 Biosafety Law. His argument rested on the premise that "life begins with and at fertilization": because Article 5 of Brazil's Constitution guarantees "to Brazilians and to foreigners' resident in Brazil the inviolable right to life, liberty, equality, safety and property," the Biosafety Law was unconstitutional by permitting the destruction of a human embryo.

The unconstitutionality action was heard by the Federal Supreme Court, which for the first time held a public hearing at which scientists for and against research with embryonic SCs could set out their arguments. At first, the controversy hinged on the definition of life: whether or not this type of embryo is a human life.

Now, clearly it is a **form** of human life, just as a fetus, a new-born baby or an elderly person are also different forms of human life. The real question is: "What forms of human life do we allow ourselves to violate?". It is fundamental to remember that the "life" mentioned in Brazil's constitution is already legally violated in certain situations: for example, in Brazil, a person with definitive loss of brain function (brain death) is considered to be dead, despite the fact that their heart is still beating, and their other organs functioning. That is an arbitrary, pragmatic decision, which facilitates organ transplants. However, it is not shared by other peoples, who consider someone dead only after their vital organs have stopped functioning.

So what about the other extreme of human life, during development of the embryo? Abortion is prohibited on the basis that it is unacceptable to destroy a fetus. In Brazil, however, if that same fetus results from a rape or represents a threat to the mother's life, it becomes a form of human life that legally can be violated; abortion is legal in such cases, under Article 128 of the Brazilian Penal Code.

As regards to embryonic SCs, the embryo in question has only about hundred cells and is frozen in a laboratory. What was being discussed at that time was in what circumstances an embryo prior to implantation — that is, one not yet implanted in the uterus — was a form of human life that can be violated.

It is important to note that, when assisted reproduction techniques were accepted in 1978, that also entailed — perhaps unwitting — acceptance of the fact that embryos, that form of human life, would be destroyed. Yes, for nearly 30 years this medical procedure has been producing surplus human embryos ("test-tube babies") the world over, which are not used for reproductive purposes and end up being frozen or simply discarded — and we have gone along with that fact quite untroubled. Why only now — when these embryos, lying forgotten in freezers, can help us to understand human biology better and find new treatments for diseases (that is, to foster more life) — has it become unacceptable to destroy them?

It was convenient to ignore the surplus embryos from assisted reproduction, because after all that technique allowed thousands of couples with fertility problems to fulfil their dream of having children. Meanwhile, the use of embryonic SCs, which could treat a heart attack or help paralytics recover the movement in their legs, is still restricted to laboratory animals. The day those cells are actually used in human patients, it will perhaps be more difficult to prohibit the therapeutic use of embryos no longer wanted by their biological parents.

In May 2008, the Supreme Court finally disallowed the unconstitutionality action, thus definitively establishing the legality of the 2005 Biosafety Law. In addition to permitting research with embryonic SCs, the law made it clear that Brazil is a lay country with a modern scientific development policy in line with those of the most developed countries. It entered the group of countries, comprising the USA, United Kingdom, France, Israel, Switzerland, China, Japan, Singapore and Australia, which invest in research with all kinds of SCs, including embryonic stem cells.

## 5.2 The Long Road to Human Embryonic Stem Cells in Brazil

I started working with mouse embryonic SCs in 1992, while studying for my PhD in the United States. When I entered Sao Paulo University in 1996, I received a grant from the Sao Paulo State research finance foundation (FAPESP) to assemble a laboratory and a team to work with these cells in Brazil. By 2000, using mouse embryos, we had established five lineages of embryonic SCs — USP-1, -2 -4, -5 and -6 — and had gained considerable experience in cultivating these cells.

In spite of the enthusiasm when the first human embryonic SCs appeared in 1998, we waited until the Biosafety Law was approved in 2005 before starting to work with these cells. We managed to import some lineages of human embryonic SCs from Harvard and learned how they differed from mouse embryonic SCs, so that shortly afterwards we managed to get them to multiply in the laboratory. However, we were depending on human embryonic SC lineages from groups outside Brazil, which are generally made available conditional on a series of restrictions as to what products and patents can be developed from them. It is one thing to collaborate with researchers in other countries (which must be done if we are to accelerate scientific development); it is quite another to depend on them. Now, if Brazil wanted to invest seriously in cell therapy, it had to master all the processes: we could not depend on cells from abroad.

Also in 2005, almost in concert with the approval of the Biosafety Law, the ministries of health and of science and technology joined hands to finance research with all types of SCs in Brazil. Under the newly approved law, our group proposed to develop Brazilian lineages of human embryonic SCs from the embryos that had been donated for research, and for which we received funding.

What does it mean to develop 'lineages of human embryonic SCs'?

It means removing the 50–100 cells of the embryonic bud from human blastocysts and enabling them to adapt to laboratory culture conditions, multiply and develop into millions of cells without differentiating, that is, maintaining their pluripotency and their enormous versatility (Figure 5.6a). Once a cell lineage is established (the cells have adapted to laboratory growth conditions), they can be split up and distributed to be used by several research groups. The development of new lineages of embryonic SCs involves different competences ranging from cell and molecular biology through to human fertilization and embryology. We had the support of researchers at Rio de Janeiro Federal University (UFRJ), who introduced us to groups in San Diego, California, with considerable experience in isolating the embryonic bud containing the potential human embryonic SCs from blastocysts. Our strategy was to bring some of these researchers from the USA to São Paulo, to teach us here, under our own laboratory conditions, how to perform the delicate first stages of adapting the cells taken from the blastocyst to

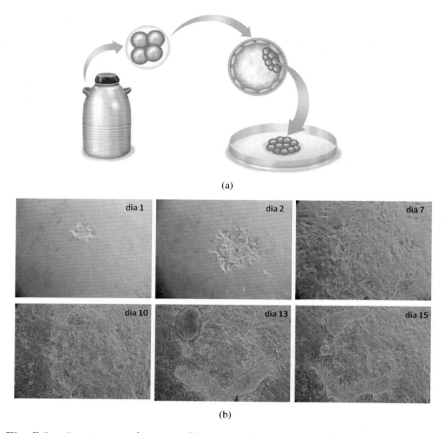

(a)

(b)

**Fig. 5.6.** Creating new lineages of human embryonic stem cells. (a) Embryos produced by *in vitro* fertilization, frozen and donated for research, are thawed and cultured to the blastocyst stage. The inner cell mass is isolated and placed in a culture flask. (b) Inner cell mass of the "embA" embryo multiplying and giving rise to the BR-1 human embryonic stem cell lineage. After 13 days' culture, regions of cells with the appearance of embryonic stem cells can be identified (dotted circles). After three weeks' culture, colonies of pure embryonic stem cells can be seen.

the laboratory environment. We recognized that human embryos are not trivial biological material and we wanted to minimize any loss of embryos during our learning process. That is why we needed to bring in researchers to pass on their knowledge.

In embryology, we received fundamental collaboration from two human reproduction clinics in São Paulo. Human reproduction clinics are the usual guardians of the embryos donated for research, and these two groups kindly placed their "rest" embryos at our disposal. All the human embryos we had available for research had been frozen while still at the 4- to 8-cell stage. However, the embryonic bud cells from which embryonic SC lineages are obtained do not emerge until later, in the blastocyst, which is a 100–120 cell embryo. Thus, the first stage of our experiment involved thawing the donated embryos and culturing them in the laboratory, so that they would grow and reach the blastocyst stage. Human embryos are extremely fragile and sensitive to the slightest variation in culture conditions. For that reason, this whole stage was carried out in the human reproduction clinics, which specialize in these processes.

The difficulties were many. Embryos donated for research are generally of poor quality — the best are used by the couple wanting a baby. Accordingly, fewer than 15% of the available embryos survived thawing and managed to develop into blastocysts. We isolated the embryonic bud from each blastocyst and transferred them to culture flasks with different culture media, with the hope that the cells would adapt to them.

This involved two years of a great deal of work and a lot of disappointing results. The majority of embryonic buds did not survive the transition and, the following day, we would find all the cells dead. At other times, the cells would start to grow, but two weeks later would stop and begin to die. We changed one reagent after another and started all over again. The team had no weekends or holidays; the cells needed caring for every day.

Until in August 2008, while experimenting with our most sophisticated culture medium, we managed to get one embryonic bud — from the embryo denoted "embA" — to grow for more than two weeks to form a cluster of cells that really looked like embryonic SCs (Figure 5.6b). Yes, we know what the different cell types "look like": whether they are fibroblasts, blood cells or neurons, each one has its own particular appearance. The cells that showed up on that plate looked very characteristically like human embryonic SCs.

Those cells went on multiplying and passed all the tests for pluripotency, including forming tumors in mice. So it was that, in October 2008, we announced we had established the first lineage of Brazilian embryonic SCs, which we called BR-1 in tribute to the federal government, which had financed out research. Establishing that lineage gave Brazil the autonomy to develop therapies with its own embryonic SCs.

Since then, in partnership with the UFRJ and with federal government funding, we have set up the National Laboratory of Embryonic Stem Cell (LaNCE), whose mission is to foster research and therapies with pluripotent SCs in Brazil. To do so, we have established more embryonic SC lineages, made cells available to our scientific community, trained researchers in how to culture the cells, and developed techniques and reagents to produce the cells on a large scale. After all, if we want one day to be able to regenerate a human organ with cells derived from embryonic SCs, we are going to need them in large quantities.

By 2012, the LaNCE-USP had already established four more embryonic SC lineages (BR-2, -4, -5 and -6). Meanwhile, the LaNCE-UFRJ had developed a new culture medium which made the cells multiply faster and managed to culture them in large flasks, thus significantly increasing cell production. In addition, the two laboratories trained Brazilian research groups, which are now studying the use of embryonic SCs in animal models to treat heart diseases, diabetes, Parkinson's disease and spinal cord injury. When the results of this research are good enough to begin trials in patients, the two LaNCE laboratories will have attained all the infrastructure necessary to produce embryonic SCs with the technical and health quality necessary for use in humans.

## 5.3 Clinical Trials with Human Embryonic Stem Cells

Nonetheless, until 2010 — in spite of all the enthusiasm over therapies using embryonic SCs — no clinical trials were conducted using such cells as a form of treatment. Why not? Well, before starting clinical trials where cells derived from embryonic SCs are actually injected into human subjects, some fundamental issues must be resolved. The first has to do with how safe such cells are. While on the one hand the great versatility of embryonic SCs is an advantage (various types of tissue can be produced from them), on the other this pluripotency represents a risk.

As already mentioned, before using embryonic SCs therapeutically in a transplant, we must first discover *in vitro*, in the laboratory, how to guide their specialization into a specific type of tissue. For example, we have to be sure that among the neurons generated from embryonic SCs there are no other types of cells present which, when transplanted into a patient with Parkinson's disease, could produce a brain tumor. Therefore, one major challenge in embryonic SC therapies is to develop efficient differentiation protocols that produce lineages containing no undifferentiated cells that might produce tumors. The literature contains a number of publications describing these protocols and testing the efficacy and safety of the differentiated cells in animal models.

A second and extremely important question is the compatibility between embryonic SCs and patient. In any transplant, donor and recipient must be compatible or the organ will be rejected (note that, with adult SCs, therapies are usually developed using the patient's own cells, so the problem of rejection does not arise). Should the same occur with a transplant of embryonic SCs?

In fact, we still do not know for sure whether, or to what extent, tissues derived from embryonic SCs will suffer rejection by patients' immune system. That is a potential issue that may make it more difficult to use embryonic SCs in therapy.

To conclude, in animal models, embryonic SCs have substantial therapeutic effect on various different diseases; but what will happen when we start human trials?

Early in 2009, approval was granted in the USA for the first clinical trial with cells produced from embryonic SCs: nervous system cells used to treat spinal cord injury were developed by the company Geron. Working with one of the first lineages of human embryonic SCs, called H9, they developed a method to turn them into **oligodendrocytes** (a specific type of cell from the nervous system) which, when injected into mice with spinal cord injury, led to significant recovery of movement in the animals.

In 2010, Geron started human trials, and by the end of 2011 had injected the oligodendrocytes produced from human embryonic SCs in 4 patients with spinal cord injury, without any adverse effects after three months of treatment. In 2015 the study was still underway, but now conducted by Asterias Biotherapeutics. Also in 2010, another USA firm, ACT (now Osaca Therapeutics), received permission to start trials in patients with macular degeneration, using retina cells produced from embryonic SCs. By 2014, after almost two years of the transplant, no tumors or immune rejection had been observed in 18 patients treated; and improvement of vision was reported in most of the treated eyes.[20] Notice that with the small number of patients treated in this phase I trial one cannot conclude that this stem cell therapy is efficacious — nevertheless, the results are encouraging.

Then, in 2014 the company Viacyte, in the USA, initiated a clinical trial using insulin-producing cells derived from human embryonic SCs to treat type I diabetes. Their approach was to pack the cells in a small capsule made of a semi-permeable membrane, and implant it under the skin. This way, the insulin producing cells are protected from an eventual attack from the immune system of the patient, and are still able to sense the levels of glucose in the blood to produce insulin according to the physiological needs of the patient.

These first studies were designed mainly to evaluate whether the procedures are safe; that is, whether the transplanted cells will behave as expected or produce some type of tumor or unexpected side-effect, or rejection of the transplanted cells. For this, the patients will have to be monitored for at least two years. Since until 2015 early results demonstrated no side effects, the expectation is that the field will advance faster and other clinical trials to treat Parkinson and heart disease with embryonic SCs will also start in the next few years.

## 5.4 Personalized Cell Therapy

Although no-one knows yet how important the issue of compatibility between recipient and tissues derived from embryonic SCs will turn out to be, one solution would be to set up a large bank of such cells, each line derived from a different embryo, to make it possible to find at least one compatible with the prospective recipient. In some countries work is being performed to achieve this.

However, in the past 40 years, bone marrow and cord blood banks have shown just how difficult it is for most patients to find compatible donors. One alternative would be to create embryonic SCs that are genetically identical to the recipient.

### Human Cloning

The first strategy used for this purpose was **cloning**, a form of asexual reproduction (not involving ovules and spermatozoa) that produces individuals genetically identical to the original.

Some living beings multiply naturally by cloning. A bacterium, a living being made up of a single cell, divides to produce two genetically identical descendants. Some plants reproduce from cuttings, pieces of themselves, to produce other complete plants that are genetically identical to the original plant — in a process called cloning.

Mammals reproduce sexually by bringing an ovule together with a spermatozoon to generate a new genome and thus a new, genetically unique individual, a mix of the maternal and paternal genomes. However, if all our cells contain a complete genome, wouldn't it be possible to produce a being genetically identical to ourselves — a clone — from any single cell of our body?

Scientists managed to clone a mammal — the sheep, Dolly — for the first time in 1996.[21] The strategy of cloning is to transfer the nucleus of a cell from the individual to be cloned, which contains the complete genome of that individual, into an ovule whose nucleus has been destroyed (Figure 5.7). When the resulting embryo starts to develop, it will follow the instructions contained in the replacement genome, giving rise to an individual genetically identical to the cell donor — a clone. This process is highly inefficient in mammals and, before a normal clone is born successfully, a series of malformed animals will abort or die shortly after birth. Nonetheless, several mammal species have now been cloned, including mice, cattle, horses, dogs, cats and even monkeys.

Cloning of humans is prohibited in all countries, including Brazil: besides the legal, psychological and social considerations, it would be completely unethical to allow such a disastrous reproduction technique in humans. However, using the same nuclear transfer strategy as used in cloning, it is possible to produce embryonic SCs genetically identical to those of the recipient.

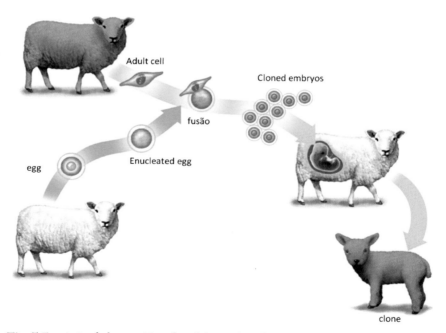

**Fig. 5.7.**   Animal cloning. Transfer of the nucleus from an adult cell to an ovule whose nucleus has been destroyed. The new nucleus is reprogrammed by the ovule and begins to behave as a zygote. The resulting embryo developed to give rise to a clone, a sheep that is genetically identical to the donor sheep from which the adult cell was taken.

What is known as **therapeutic cloning** starts with the transfer of the nucleus of any cell from the patient to an ovum in which the normal nucleus has been removed, thus creating an embryo genetically identical to the patient (Figure 5.8a). That embryo is cultured in the laboratory and when it reaches the blastocyst stage, instead of transferring it to a uterus — which would constitute reproductive cloning, embryonic SCs are taken from the embryo. As these cells are genetically identical to the recipient, in theory, no tissue derived from them will suffer rejection when transplanted.

By 2013, therapeutic cloning — or "nuclear transfer," as some prefer to call it, so as to steer clear of the controversial term **cloning** in connection with humans — had been done in mice and monkeys, but had not yet been successful in humans. Since the technique is not very efficient, it calls for large numbers of ova and human ova are not easy

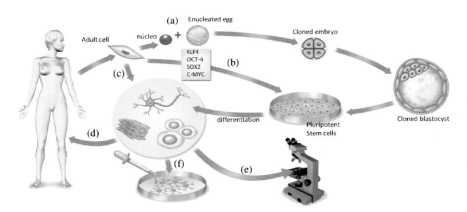

**Fig. 5.8.** Cell reprogramming. There are three ways to change the identity of an adult cell: (a) Therapeutic cloning: by inserting the nucleus of a skin cell into an ovule, a cloned embryo is created, which can be cultured in the laboratory to the blastocyst stage. The inner cell mass will give rise to embryonic stem cells genetically identical to the recipient's and which can now differentiate to become the cell type needed to treat their disease. (b) Induced pluripotent stem cells: the same skin cell can be reprogrammed by inserting the oct4, klf4, c-myc and sox-2 factors, which cause it to regress to the pluripotent cell stage. (c) Direct reprogramming: by introducing other factors into the skin cell, it can be induced to turn directly into the desired cell type, without having to go through the embryonic stem cell stage. The differentiated cells produced by this method can be used (d) for transplanting, (e) for research or (f) for trialing new drugs.

to obtain. Then in June 2013 a group led by scientists in Oregon, USA, reported the generation of lines of pluripotent human SCs by nuclear transfer.[22] By optimizing different steps of the process, the researchers were able to reprogram skin cells of a patient with a neurological disease into pluripotent SCs! Soon another group reported the generation of nuclear-transfer pluripotent SCs from a diabetic patient, consolidating the technique.[23] We then became able to produce personalized pluripotent SCs! Well, not exactly then — six years before, in 2007, a group in Japan had developed a different approach that had already revolutionized the field.

## Induced Pluripotent Stem Cells

By 2006, while some groups were investing in nuclear transfer as one way of producing human embryonic SCs genetically identical to the recipient, another much simpler strategy was being developed in Japan.[24] In nuclear transfer, when the nucleus of an adult cell is introduced into an ovum, unknown factors in that ovum act on the cell's genome, reprogramming it to behave like the genome of an early zygote, the first cell of the embryo, resulting from the fusion of ovum and spermatozoon.

This means that, while the genome of a skin cell (fibroblast), for instance, has only fibroblast genes activated, the reprogramming of that nucleus will "turn on" another set of genes, those specific to a zygote. From then on, that "zygote" will begin embryonic development until it gives rise to a blastocyst, from where embryonic SCs can be obtained.

While some research groups tried to identify these factors in the ovules that caused this reprogramming of the adult genome, a group at Kyoto University decided to investigate what it is that makes embryonic SCs different from other types of cell. What set of genes was active in embryonic SCs that was not active in adult cells?

They examined the different types of cells and selected 24 genes that were turned on in embryonic SCs and off in other cell types; if they could manage somehow to reactivate these genes in an adult cell, would it turn into an embryonic SC?

One way of activating genes in a cell is to insert copies of those genes into it — and this can be done relatively easily, thanks to the recombinant DNA techniques mentioned earlier. Well then, when copies of those 24 genes were introduced into mouse fibroblasts, after a few days, these cells turned into cells that strongly resembled embryonic SCs and were also pluripotent — these cells were able to give rise to any tissue type.

Now, were all those 24 genes really necessary to bring about this metamorphosis?

The Japanese researchers went further and using various combinations of those factors, arrived at a set of 4 genes that was necessary and sufficient to induce the transformation of a skin cell into a pluripotent SC. These were the *Oct4*, *c-Myc*, *Klf4* and *Sox2* genes, which are very active in embryonic SCs.

These genes exist in all our cells, but are turned off in differentiated cells. When active copies of the four genes are inserted into a fibroblast, they will produce the respective proteins characteristic of embryonic SCs in that cell. No one knows quite how yet, but these four proteins are sufficient to reprogramme the genome of the adult cell, activating or deactivating a series of other genes, thus causing the cell to go back in time and revert to the pluripotent cell stage, equivalent to an embryonic SC.

As these cells were not originally pluripotent — they were induced by the four genes to become that versatile — the Japanese group baptized them **induced pluripotent stem cells** (iPSCs). The following year, using the same set of four genes, that same group managed to produce iPSCs from human fibroblasts.[25] Now these pluripotent SCs, which are genetically identical to the recipient, can be differentiated into the specific cell type needed to treat that individual's disease and theoretically with no risk of the cells being rejected by the immune system (Figure 5.8b).

Induced PSCs have revolutionized stem cell research, and this simpler method for obtaining pluripotent SCs was quickly adopted by dozens of laboratories around the world. Some went as far as to say that, with iPSCs there was no more need to work with controversial embryonic SCs. After all, they now had an easier method for obtaining pluripotent SCs, which involved neither ovules nor destroying human embryos.

However, it is still too early to say whether iPSCs are better than, or identical to, embryonic SCs, and such comparisons are the subject of research in a number of laboratories around the world. In terms of their therapeutic use, iPSCs entail the same risk as embryonic SCs of producing tumors; that risk is inherent in any pluripotent SC. Further, to turn differentiated cells into pluripotent SCs, exact copies of those four genes were introduced into their genome — and that represents an additional risk of those cells misbehaving in unexpected manners. The great advantage of iPSCs over embryonic SCs is that they are genetically identical to the recipient and so, in theory, will not produce an immune response when transplanted.

Note that, although therapeutic cloning or iPSCs may perhaps solve the problem of compatibility between pluripotent SCs and recipient, these strategies cannot be used in individuals with genetic diseases such as hemophilia, cystic fibrosis or muscular dystrophy. These diseases are caused by mutations, by defects in specific genes. Pluripotent SCs produced from such patients' cells would also carry the genetic defect and, accordingly, would be unsuitable to produce healthy tissues for transplantation. Unless we were able to correct the genetic defect of the cells in the lab… and believe it or not, this is becoming a real possibility!

Think about the challenge: to find and correct the single wrong "letter" among the 3 billion letters that compose our genome! Well, in the last 5 years, sophisticated and yet simple to use techniques were developed that allow us to do exactly that in cells. And thus, these genetically corrected iPSCs can generate healthy patient-specific tissues that may one day be used for therapy in individuals with genetic diseases.

So, let's imagine that one day the safety problem with iPSCs will be solved and the defective genes in cells from people with genetic diseases will be fixed and healthy tissues produced for transplanting (which should happen before too long). Then will personalized cell therapy actually be feasible?

Look at it this way: before the oligodendrocytes produced from H9 embryonic SCs gained approval for use in humans, the company Geron had to test them in more than 2,000 animals to demonstrate that they were safe. If they now want to use another embryonic SC lineage — our BR-1, for example — to produce those same nervous system cells, they cannot simply apply the same methodology to differentiate them into oligodendrocytes as developed for the H9 lineage and assume that the BR-1 embryonic SCs will behave exactly the same way and produce safe tissues. They will have to test the oligodendrocytes produced from the BR-1 cells in animal models in order to demonstrate that they too are safe (that they do not produce tumors).

So, if personalized cell therapy is expected to produce pluripotent SCs genetically identical to each patient, then in principle each therapy would have to be tested in animal models before being applied to the patient. Imagine the time it would take to develop each therapy: to produce the patient's iPSCs; in the case of genetic disease, to correct the gene defect in the cells; to differentiate them into the desired tissue; to test them in animal models to prove they are safe — and the costs involved ... Although the technical and scientific aspects of personalized cell therapy with iPSCs are complex, they can be worked out. What to me seems more of an impediment is the procedure's economic feasibility for treating a large number of patients.

## Direct Reprogramming

If a cell can be reprogrammed, why not reprogram it directly to the desired cell type, without having to go back to the embryonic SC stage?

If the nucleus of a skin cell contains a complete genome, then it also has all the information necessary to become, for instance, a neuron. Now, perhaps it is possible to turn on that programming to become a neuron without having to go all the way back through the process to the embryonic stage, then to induce neural differentiation (Figure 5.8c). Theoretically, that would be quicker and would produce safer cell populations, because it would avoid going through the pluripotent cell stage, which can form tumors.

The challenge was to discover which genes to activate in order to directly reprogram a skin cell into a neuron or a cardiomyocyte and so on. The best way of discovering that was to study the specific genes that are activated during the development of each tissue type in the embryo. In this way, putting that information together with a good measure of trial and error, in February 2010 a group at Stanford University, in the USA, described the direct reprogramming of mouse fibroblasts into neurons.[26] They showed that when expression of the *Brn2*, *Ascl1* and *Myt1l* genes — which are active exclusively in nervous system cells — is induced in fibroblasts, in 12 days those skin cells began to behave as neurons, including having the capacity to transmit electric signals. In August 2011 the same group described the direct reprogramming of human fibroblasts into neurons using the same set of genes.[27]

From then on, a series of other studies were published describing the direct reprogramming of fibroblasts into various different cell types. For each type, expression of a specific set of genes has to be induced, according to the cell type desired. Thus, if the *Gata4*, *Mef2c* and *Tbx5* genes are activated in fibroblasts, these will turn into cardiomyocytes (heart muscle cells).[28] However, if those same fibroblasts are to be reprogrammed into liver cells, then the *Gata4*, *Hnf1a* and *Foxa3* genes must be activated, and the *p19Arf* gene turned off.[29] If the intention is to produce blood cells, *OCT4* expression has to be activated in the fibroblasts, and cells must be placed in a blood cell culture medium.[30]

This cell reprogramming can be even more sophisticated and generate even more specialized cell types from fibroblasts. For example: there are many different types of neurons in our nervous system and patients will have different neurological diseases depending on which subgroup of neurons is ailing. Parkinson's disease and amyotrophic lateral sclerosis

(ALS) are neurodegenerative diseases. In each one, however, a different type of neurons is affected and, accordingly, the cell therapy for each one must involve producing that specific type of cell.

Parkinson's disease is caused by loss of highly-specialized, dopaminergic neurons, which produce the neurotransmitter dopamine. Dopamine deficiency causes patients to develop tremors and involuntary movements. In order to produce that cell type from pluripotent SCs, first these would have to be differentiated into immature neurons and then be made to develop into dopaminergic neurons. With direct reprogramming, it was discovered that activating the *Mash1*, *Nurr1* and *Lmx1a* genes in fibroblasts transforms those skin cells directly into dopamine-producing neurons.[31]

However, ALS is a disease characterized by the degeneration of motor neurons, which transmit nerve impulses to muscles and control voluntary movements. Loss of those neurons results in loss of muscle stimulation and the muscles atrophy. Accordingly, to reprogram a fibroblast into a motor neuron, not only the *Ascl1*, *Brn2* and *Myt1l* genes have to be activated, as necessary to produce neurons, but also the *Lhx3*, *Hb9*, *Isl1* and *Ngn2* genes.[32]

The prospects for direct cell reprogramming are fascinating. It has the potential to enable us to produce cells and tissues for therapy more safely and effectively. However, much has to be learned about the gene circuits involved in the development of each cell type, about the sequence of events that transforms that single cell resulting from fertilization into an individual as complex as a human being.

## 5.5 Studying Human Biology with Pluripotent Stem Cells

To me, one of the most fascinating questions in biology is: how do we go from one day being a single cell to the trillions of cells we are now? How are those initially identical cells of the embryo reorganized to multiply and specialize to give rise to the more than 200 different tissue types in our bodies? And how can we study this process in humans if it takes place inside the uterus?

One way is to use embryonic SCs. Remember that these can be induced to differentiate in culture flasks in the laboratory: for example, when we cause them to become muscle or neuron cells in a culture vessel in the laboratory, they are recapitulating the process of differentiation that they would normally undergo in the human embryo, which enables us to observe and try to understand all the events necessary for that transformation. That is, the *in vitro* tissue specialization of pluripotent SCs in the laboratory are a valid experimental model of human embryo development.

Of course, in the culture flask, the cells are not going to grow into a baby, nor a fetus or embryo; but rather will result in specialized cells, and even tissues with some degree of organization, as occurs during the development of a human embryo. Then, we will be able to study in detail the sequence of events that leads an undifferentiated cell to become blood, liver and so on. Which genes are activated? Which turned off? In what order?

That is basic research designed to produce pure knowledge, without necessarily any view to application. This type of research, as the name suggests, is the basis for developing applied research, the intention of which is to use pure knowledge to develop some new method or product. Now, when we unveil the mechanisms involved in the ability of embryonic SCs to turn into any cell type, we are learning about the biology of the human being — about how we work. In the long term, this basic knowledge can yield great benefits to human health.

## Induced Pluripotent Stem Cells as Models of Genetic Diseases

On the other hand, induced pluripotent stem cells (iPSCs) generated from cells of patients with genetic diseases can help us in understanding the basic mechanisms behind such diseases. Remember that in the case of individuals with genetic diseases, their iPSCs would have the same genetic defect and, therefore, be unable to generate healthy tissues to treat their disease without some genetic modifications to correct the disease. However, iPSCs, can be used as a tool for studying aspects of the disease in the laboratory.

Let's go back to the examples of Parkinson's disease and Amyotrophic lateral sclerosis (ALS). Various neurological alterations are known to take place in patients, but how can the dopaminergic neurons of Parkinson's or the motor neurons of ALS be accessed in order to study them in detail?

Some patients donate their brains for research, but such material is scarce and not always suitable for this kind of research. After all, the idea is to monitor the life of those neurons from their origin from undifferentiated cells through to when they start to degenerate.

However, since we now know how to reprogramme skin cells, we can take small tissue biopsies from the patients, multiply the fibroblasts in the laboratory, produce iPSCs, differentiated them into nerve cells resulting in an almost inexhaustible source of neurons for all these neurological diseases.

Now, will these cells from the patients reproduce the disease in the culture flasks? How does a culture of neurons with Parkinson's behave? Or a culture of heart muscle cells with arrhythmia? Actually, clinical diagnosis of Parkinson's disease or arrhythmia is relatively simple in patients, but will it be possible to identify these symptoms in cells in culture?

The first studies describing iPSCs produced from individuals with genetic diseases emerged in 2008, including one with iPSCs derived from a patient with ALS (the neurodegenerative disease caused by motor neuron defects). In this patient, the disease resulted from a mutation in the *SOD1* gene, and the pluripotent cells produced contained the same mutation.[33] In order to understand how genetic defect causes the damage to motor neurons, the researchers produced nerve cells from the patient's iPSCs, and also from iPSCs from normal individuals. Comparisons between the two groups of motor neurons may yield new knowledge about the mechanisms of the disease.

Another example is a study of long QT syndrome (LQTS)[34] a genetic heart disease characterized by tachycardia and arrhythmias and sudden death from heart failure (QT being a measure on the heartbeat curve). The researchers not only produced iPSCs from a patient with the syndrome, but produced cardiomyocytes (heart muscle cells) from them. In culture flasks, these cells form clusters that contract rhythmically, as if they were part of a heart.

[See again these cells contracting in the video at http://www.ib.usp.br/lance.usp/bookct/video3].

This study went further and, by comparing the cardiomyocytes from a patient with LQTS with cardiomyocytes from a normal individual, the researchers showed that in culture, the cells with the genetic defect also displayed arrhythmia and other electrical pulse alterations at each contraction.

How was that? Did they run an electrocardiogram on the cells in culture?! Yes, using microelectrodes stuck to the culture flasks, it was possible to evaluate a number of heartbeat parameters in that cluster of cardiomyocytes.

In addition, when drugs known to worsen the clinical condition of patients with LQTS are placed in a culture medium of diseased cardiomyocytes, they also worsen the condition of those cells. That is, the researchers managed to demonstrate that those cells were in fact behaving in culture just like the cells in the patient's heart.

What can be done with the diseased cardiomyocytes that have been produced?

If they deteriorate with drugs that are harmful to the patient, it is reasonable to imagine that they will respond favorably to drugs that improve the heart condition of individuals with LQTS. Numerous known drugs and other new ones can be tested on the diseased cardiomyocytes to see which of them normalize the cells' electrocardiogram. Such substances will be strong candidates for trials in humans.

This strategy is being used to understand the molecular mechanisms behind a number of diseases and to identify potential new drugs to treat them. Neurons produced from the iPSCs of children with forms of autism or with Alzheimer's or even Parkinson's disease or other genetic diseases of the nervous system can now be studied in detail in the laboratory. What metabolic pathways are altered in each group of these cells? What drugs are able to reverse the cell defect? The expectation is that, using iPSCs as a cell model for various diseases, it would be possible to find alternative therapies for them quickly.

## 5.6 Pluripotent Stem Cells for Drug Development

Well, we have seen that pluripotent stem cells, embryonic SCs or iPSCs can be used for cell therapy and for basic research. The third application of these cells is in developing new drugs.

This is a very lengthy process starting with trials of literally millions of compounds (Figure 5.9). Of these, a few thousands will be identified as showing promise for the desired application, and their formulation

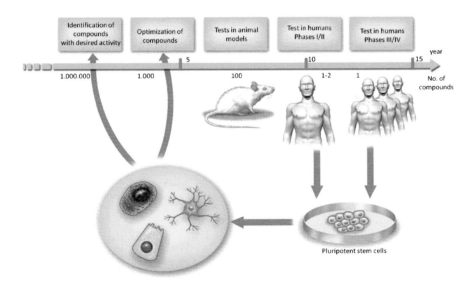

**Fig. 5.9.** Development of new drugs. Timeline showing the years and number of compounds trialed in order to bring a new drug to the market. Pluripotent stem cells are used to expedite this process by identifying earlier which compounds potentially offer better activity and lower toxicity in humans.

will be improved to give rise to a few hundred compounds. These, in turn will be tested in animal models, until finally one or two will be identified as worth trialing in humans. To reach this vital stage of clinical trialing takes around 10 years and consumes an enormous amount of money. There will then be yet another 5 years of clinical trials, if everything goes well, before a new drug emerges on the market (see the box "From basic research to clinical trials," on page 25).

One major problem facing the pharmaceutical industry is that, after years of investment and research, some 90% of new drugs tested in clinical trials are not approved, very often because they cause heart arrhythmia — one of the commonest reasons. It is very important that pharmacists identify any drugs that cause this kind of side effect as early as possible, preferably before they are trialed in human subjects. This will avoid exposing people to the risk of arrhythmia, and companies can save time and money in developing new drugs.

Drugs can be tested in animal models and this is done, mainly in mice. However, mouse hearts, which beat 600 times a minute, are very different from ours, which beat 80 times a minute. Important differences in size, blood pressure, susceptibility to heart attack and arrhythmias limit the predictive value of cardiac toxicity tests in such models. So, at present, there is no effective way of testing whether or not a new drug has cardiac side effects before it is trialed in humans.

But what if that drug could be tested on human tissues, more specifically, in human cardiomyocytes?

Just as arrhythmia can be measured in cardiomyocytes from patients with long QT syndrome, so heart cells can now be produced from embryonic SCs and treated with the new drug to see whether, in those conditions, that human heart tissue develops arrhythmia. Even if the drug does not induce the effect in those cells, this does not necessarily mean it will not cause arrhythmia in people. On the other hand, though, if it does demonstrate any toxic effect on cells in culture, it will not even go on to be trialed in humans.

Now, as any type of adult cell can be produced from embryonic SCs, the same strategy can be applied to evaluate the toxic effects of a new drug on the nervous system or on the liver. Neurons or liver cells can be produced from embryonic SCs and used to test these drugs in the laboratory before they are administered in humans. This application of embryonic SCs is already a fact. There are firms whose business plan is to produce cardiomyocytes, neurons and hepatocytes (liver cells) derived from human embryonic SCs and to sell those cells to be used for trialing new drugs. They are already being used in the early stages of screening to identify, from among millions of compounds, those that are of interest for the desired application and also for safety testing to complement the trials with animal models. Thus, even before these cell types are used in transplants and organ renewal, they are already useful in developing new drugs.

## Forecasting Drug Response in Different Populations

You probably already know that certain drugs produce different responses in different people. Antidepressants are one example. There are various types of such drugs and each one will function for only a certain subgroup of patients. Why is that?

There are at least two possible explanations: depression may have differing basic causes, and each drug may manage to treat only one specific type of the disease; each person may metabolize drugs differently, causing them to have varying effects on the cell system being tested.

So, what determines how drugs are metabolized?

Several factors are involved, including age, state of health and diet — but, arguably the most important of all — your genes! Yes, your genes also influence how a drug is absorbed, processed, transported and eliminated. Small variations in genes produce small variations in each of these processes — and, taken together, these variations will cause a drug to perform well therapeutically in one person and not in another. The study of how variations in genes, in your genome, affect your response to drugs is called **pharmacogenetics**.

Pharmacogenetics has its success stories. A case in point is fluoxetine, a type of antidepressant better known by its brand name, Prozac.

In order to ascertain what dose of fluoxetine a patient should take, the doctor may ask for an analysis of their *CYP* genes. Depending on what gene variants the patient has, they will metabolize the drug faster or more slowly. Fast metabolizers will need higher doses of the drug to maintain an appropriate body concentration for any length of time. Slow metabolizers need lower doses of fluoxetine.

What is more serious than a drug having no effect is its being toxic. To give you an idea of the magnitude of this problem, in England, one in every 15 cases of hospital admissions is due to adverse drug reactions, while in the USA, two million patients per year suffer severe adverse reactions and, 100,000 of them will die. In the same way that our genes influence how effective the drug is for us, they also influence how toxic they are.

To sum up, drugs may simply not work or may have toxic effects on some people. How then does one decide when a new drug can be released for use on patients?

On the criteria currently used by the pharmaceutical industry, for a new drug to be approved for sale, it needs to function well in only 30 percent of the people tested. However, these drugs are trialed in general populations in Europe and the USA, and then sold in developing countries before it is known how effective or safe they will be for these other populations.

Why would a drug behave differently in other countries? This is because, although each person in a given country has a unique genome, people from the same population tend to have the same genetic variations in their genomes. We have seen that genome variations influence drug response and that some of these genetic variations may be more frequent in one population than another, making such a population more susceptible, on average, to a drug's adverse effects.

Take the example of BiDil, a medicine for congestive heart failure. To begin with, this drug was tested on a population of white North Americans and proved ineffective. In a second clinical study, conducted exclusively with North Americans of African origin, BiDil was found to reduce the likelihood of death following heart insufficiency significantly in that group. Accordingly, in 2005, BiDil became the first drug approved for use in a specific population — in this case, in patients self-identified as 'black'. Individuals in that population have genetic variants that cause them to respond much better to this drug than 'white' individuals.

However, if each new drug has to be tested in different populations, that would increase enormously the cost and the time taken to put new drugs on the market! This is true and is why we need alternatives to clinical trials. That is where iPSCs come in.

Instead of trialing a drug in various different individuals from various different populations, how about testing it on cells from various people from those populations?

Yes, we can produce iPSCs from several different people representing a specific population — be it North American, Indian, Japanese or even Brazilian — and before starting to test the drug on one of these populations, it can be tested on the population of cells to evaluate its toxicity to them. The idea then is that, with these cell libraries, it will be possible to forecast a new drug's adverse effects or variations in its efficacy in each one of these various human populations, even without testing it on the people involved.

As already mentioned, trials on cells in culture are not perfect either and we will still need to trial new drugs in humans before finally approving them for sale. However, imagine a common scenario where there are five candidate drugs in line for trialing in human patients — which should be trialed first? Trials in cells can help to identify the most promising drugs more rapidly, so that efforts can be concentrated on those.

## Forecasting Drug Response — Individualized Medicine

Although it is recognized how important genetics is to pharmacology, still little is known about what genes are involved in responses to different drugs. At present, a doctor will not know if, for example, a drug causes arrhythmia in the patient until the individual has suffered that major discomfort. It has been shown that we can use pluripotent SCs to discover new drugs and to test their effect in different populations. Within any given population, however, there will be individuals who respond well and others who respond badly to a drug.

As has been seen, populations tend to have genetic variants in common, but each individual has a unique genome and, therefore, will have their own unique response to the drug. How can this be predicted?

By once again using iPSCs. Imagine if, before prescribing a drug for a patient, it could be tested on a piece of their heart muscle in a laboratory to see if it causes arrhythmia. The difficult part would be persuading the patient to undergo a heart biopsy for that purpose ... However, if that individual were biopsied and the respective iPSCs were produced from the specimen, that would provide a limitless source of the patient's cardiomyocytes. That heart tissue could be used to test its susceptibility to the drugs in the laboratories. All that has to be done is to place the drug in the culture of cardiomyocytes derived from the patient's iPSCs and determine whether or not the cells suffer as a result.

By the same argument, neurons can be produced to test the effectiveness of an antidepressant, or liver cells to measure how fast each patient metabolizes a drug. That is to say, even before knowing what genes control the response to each drug, personalized medicine can be practiced using iPSCs to predict how a patient will respond to a given drug. Who knows, the time may come when each of us will have our iPSCs established to be tested before we are prescribed a drug.

# 6 Cell Therapy — Promise or Reality?

I hope, that by now, I have managed to demonstrate the vast prospects that exist for therapy using the various different types of stem cell, adult and pluripotent (whether embryonic SCs or iPSCs). What is more, for a number of diseases, research with embryonic SCs and most importantly with adult SCs, has now taken the major step from animal models to trials in human beings.

However, it is important to be clear that **this type of therapy is still restricted to the realm of research**. In other words, although the results of using SCs to treat these diseases are promising, no doctor can prescribe such therapy for patients. Until 2015, the only very well established treatment with SCs involved transplants of bone marrow or umbilical cord blood to treat diseases of the blood (Table 4.1, page 28).

Nonetheless, unfortunately, in many countries there is a large-scale clandestine trade in SC miracle treatments, which exploits the despair of patients and relatives looking for alternative therapies for diseases that today are incurable. Clinics use the Internet to announce stem-cell treatments for multiple sclerosis, spinal cord injury, ALS and even Aids, along with other diseases, exploiting loopholes in their countries' legislation. The scientific community vehemently repudiates these practices, which are unsupported experimentally, unethical and expose patients to unnecessary risks.

Some families argue that they have nothing to lose in these attempts, but they are mistaken. In 2009, an article reported the outcome of one of these illegal stem-cell treatments: at a clinic in Russia, a boy seeking treatment for his neurodegenerative disease developed multiple brain tumors.[35] Analysis of the genetic material from the tumors revealed that the cells they were made up of were genetically different from the boy's (that is, they were the cells injected at the Russian clinic). Not only that, but this examination of the tumors' DNA revealed that the cells injected into that patient came from more than one person (or fetus or embryo — how is one to know?).

In 2011, a clinic that offered stem-cell treatment in Germany was closed following the death of an 18-month-old child and another 10-year-old who had received stem-cell injections into the brain and spinal cord. The clinic had been operating since 2007, taking advantage of the lack of specific legislation on cell therapy in Germany, and had branches in two German cities. For up to €26,000, they promised to treat strokes, cerebral palsy, ALS, multiple sclerosis and other diseases.

Even in the USA, in 2012, there were at least two companies selling treatments using adult SCs (from bone marrow or fat) for a variety of conditions, including bone and cartilage regeneration, multiple sclerosis and Parkinson's disease. These companies availed themselves of the fact that the legislation on stem-cell treatments was still under construction in the USA (as in many countries; after all, this is a very new field of medicine) and marketed the miracle stem cells as if they were offering established treatments.

What is most confusing is that some of the patients treated at these clinics do in fact show some improvement, which makes it almost irresistible to want to undergo such procedures. Can it be that some of these clandestine treatments have some value?

Unfortunately, there is no way of knowing it. The clinics do not publish details of their treatments: what cells are injected and in what quantities, what scientific evidence the treatment is based on, or how patients who undergo the treatments compare with others with equivalent clinical conditions who do not. That is, are those who do show improvement just part of that fraction of patients who would improve naturally or can their progress actually be attributed to the stem-cell treatment?

These questions can only be answered if there is total transparency in how the therapies are applied and if carefully controlled clinical trials are performed. This is where I ask myself: if someone really did discover a cure for Alzheimer's disease, cerebral palsy or any of these other terrible diseases, why not follow the orthodox pathways of scientific research, publish the results and, in that way, not only gain total credibility for the procedure, but even be in the running for a Nobel prize? Why prefer to be clandestine, to sell this as a mysterious and controversial remedy? The only reason I can think of is that there must be something wrong with the therapy — either in its scientific grounding or in the results it yields.

So — caution — for now, there are no proven stem-cell treatments for any of these diseases. Therefore, at best, the treatments on offer out there should be treated as **experimental therapies** and not miracle cures — and experimental treatment should be performed only in public or private research institutions, with the approval of the respective research ethics committees and at no financial cost to the patient. We understand and sympathize completely with the suffering and anxiety of the patients and relatives who are waiting for these long-promised stem-cell treatments. First, however, it has to be established that these therapies are safe and then that they really are effective in treating these diseases, before they become established medical procedures and are made available to the public.

The clandestine trade in stem-cell therapies has reached frightening global proportions (Figure 6.1) and come to constitute "stem-cell tourism."[36] In order to formally repudiate and combat the "stem-cell merchants," the International Society for Stem Cell Research, which brings together the world's leading scientists working with stem cells, created the site "A Closer Look At Stem Cells" (http://www.closerlookatstemcells. org/), with information for the general public on the basic science and medical applications of SCs, including "Nine Things You Should Know

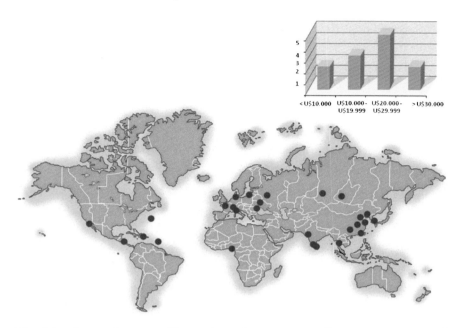

**Fig. 6.1.** Stem cell tourism. Places where clinics operate selling unproven stem cell treatments, and the prices of those treatments as of 2009 (Adapted from Ref. 33).

About Stem Cells Treatments" (http://www.closerlookatstemcells.org/stem-cells-and-medicine/nine-things-to-know-about-stem-cell-treatments) and the Patient Handbook on Stem Cell Therapies (www.closerlookatstemcells.org/patient-resources/patient-handbook-pdf). This manual serves as a guide for assessing the validity of treatments. It discusses the criteria on which research becomes medicine, how a clinical study functions and what to ask about theoretical stem cell treatments.

# 7 Outlook

Until recently, medical interventions rested on three pillars. The first was medical devices, from the simplest, such as crutches, through to sophisticated equipment, including all surgical instruments, right up to CT scanners and hemodialysis machines. Development of these devices has made use of a great deal of knowledge from physics and engineering, and a whole medical device industry has grown up to build them.

The second pillar is **drugs**, tiny chemically-synthesized molecules, such as aspirin (21 atoms) (Figure 7.1). Chemical synthesis is a well-known process — such drugs have been used for a long time — and we now control these processes so well that, after their patents have expired, these molecules can easily be produced by other groups, giving rise to what are known as generic drugs. The traditional pharmaceutical industry develops and produces these drugs.

A short time later, **biological drugs** emerged, to form the third pillar of medical intervention. These are large molecules, generally proteins ($3 \times 10^3$ atoms) or antibodies ($2.5 \times 10^4$ atoms). Because of their greater complexity, we do not know how to synthesize biological drugs chemically and they have to be produced by either human or even animal cells in culture. That is to say, they are synthesized biologically, produced from a living being (a cell), thus the name "biological drugs."

Examples of this kind of drug are insulin for diabetics, the coagulation factor IX for hemophiliacs, and the antibody Herceptin® used in treating breast cancer. Techniques for manipulating DNA have enabled

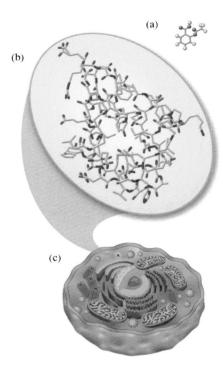

**Fig. 7.1.** Evolution of drugs: from chemical synthesis to cell therapy. (a) Small molecules, such as aspirin, can be synthesized through well-controlled chemical reactions. (b) Complex biological molecules, such as proteins and antibodies, must be synthesized by cells, from where they are purified. (c) Cells are several orders of magnitude more complex than these proteins. Accordingly, for them to serve therapeutic purposes, their production must be even more strictly controlled.

us to isolate the genes that codify each of these proteins and to insert copies of these genes into cells in culture. Those cells are then multiplied in great quantities and they synthesize and secrete the respective proteins into the culture medium, from where they can be purified to be administered to patients.

However, while chemical synthesis is a process that we can control very well, we do not have the same control over protein synthesis by cells. The same DNA of the insulin gene, when inserted into cell A, may produce a slightly different form of insulin from when it is inserted into cell B. The tiniest differences can cause a protein to be less active or to be rejected by the immune system. That is why it is so much more difficult to produce 'generics' of biological drugs (known as biosimilar drugs to underline this greater complexity). All this work with biological drugs is conducted by biotechnology companies.

We are going through a historical moment in the development of the fourth pillar of medical intervention, namely cell therapy: **cell therapy** intends to use cells or tissues as therapeutic agents. The complexity of a cell ($10^{13}$ atoms?) is several orders of magnitude greater than that of a protein. Production of cells for therapy will thus depend on other competences and, as a result, the cell therapy industry is still in the process of developing.

Cell therapy comprises basically two very different models of intervention. In the **autologous** model, cells are isolated, multiplied and produced specifically for each patient. That is the case with cord blood banks or with therapies that involve removing a small quantity of cells from the patient's body, manipulating and expanding them in a laboratory and re-infusing the resulting cells back into the patient (as carried out to treat heart attack victims with cardiac stem cells, and is intended to be done with iPSCs in personalized cell therapy). In this model, the cell therapy resembles more a medical procedure than administration of a drug. The cell product has to be manufactured individually for each case and consists of cells from the patients themselves — thus the name **authologous therapy**.

In the **allogenic** model, the intention is to use cells in the same way as drugs: a set of cells from one individual can treat several different patients. That is how the companies Asterias Biotherapeutics and Osaca Therapeutics are developing their products derived from embryonic stem cells for spinal cord injury and retina regeneration, respectively. Large batches of differentiated cells are produced from a single lineage of embryonic SCs and divided into thousands of doses to be used in different patients.

This model resembles the model used in producing chemical and biological drugs and in principle is easier to implement. While the autologous model is less high-tech and the cells must be produced to order, the allogenic model allows cells to be produced *en masse* and to be available immediately when needed, rather like a drug.

Each model has its advantages and its limitations, whether technical, financial or logistical. It remains to discover which model will be feasible for which cells and for which diseases.

Of course, these four pillars interact with one another, and products of the device, drug and biotechnology industries are fundamental to the development of the cell therapy industry. However, the latter does have its peculiarities, and for it to develop appropriately specific regulations must be put in place to protect people from hazardous treatments, but without hampering the growth of this new approach to therapeutic medicine.

In conclusion, in years to come we will be harvesting the fruits of all the basic and clinical research with the different types of SC (Figure 7.2). We will know what cells are most suited to treat each disease; the therapeutic value of other types of mesenchymal SCs, such as those from fat, umbilical cord and placenta; we will learn to isolate

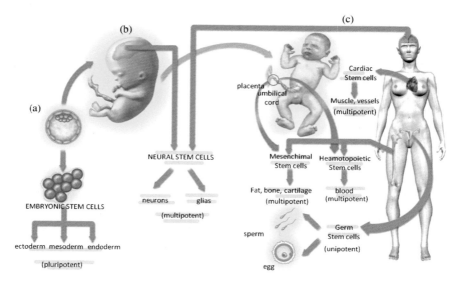

**Fig. 7.2.** Summary of types of stem cell. Different types of stem cell identified in humans: (a) embryonic; (b) fetal (c) adult.

and to multiply more types of tissue-specific SCs, such as those of the heart, the nervous system and germ cells; and we will manage to control the specialization of embryonic SCs so as to produce tissues that are safe for use in humans. In that way, finally, we will be able to see the substantial therapeutic effects observed in animals reproduced in patients, treating Parkinson's disease and diabetes or helping paralytics recover their movements.

By studying iPSCs from patients with different genetic diseases, we have managed to identify the exact mechanisms for some of these diseases and to develop more intelligent therapies for them. The possibility of testing different chemical compounds directly on patient derived iPSCs will make it easier to discover new drugs for their respective diseases. In addition, using cells derived from iPSCs to test the effects of new drugs offers the potential to minimize the risks of clinical trials in humans and to shorten timeframes for developing and marketing new drugs.

Even while we are currently unable to use these cells for therapeutic purposes, the basic knowledge of human biology acquired in stem-cell research is extremely valuable and will certainly improve quality of life. For example, by understanding what signals cause a muscle stem-cell to multiply and give rise to more muscle (as happens when you spend a lot of time at the gym), we can develop drugs that stimulate the SCs of patients with muscular atrophy to generate new tissue — without needing to inject any cells into those individuals.

Although many patients urgently need stem-cell therapies to be made available as soon as possible, the corresponding scientific development must be conducted with the utmost responsibility, so as not to expose these people to unnecessary risks. The situation is reminiscent of the process of research into heart transplantation. In 1958, the USA began to experiment with heart transplants in animal models. It took nearly 10 years of research before the scientists felt confident enough to test the procedure in humans.

Accordingly, it was not until December 1967 that a group in South Africa performed the world's first heart transplant. The second was carried out in the USA a few days later. Soon afterwards, other groups in several different countries also began to perform the procedure, so that just one year later more than 100 transplants had been performed all over the world. In 1968, in Brazil, Dr. Euryclides de Jesus Zerbini performed the world's fifth heart transplant (Brazil's first), at the Hospital das Clínicas de São Paulo. However, the overall mortality rate was 80%!

What were they to do? Stop transplanting and abandon the idea?

Without a doubt, transplantation had to be halted to make it possible to invest in improving the technique to solve the problem of high mortality. The researchers went back to their laboratories and as those transplants had been performed in research institutions and had been carefully controlled and documented, they were soon able to discover that the major problem was rejection of the organ by the patient's immune system. Once the problem had been identified, the scientists knew what to work on: discovering a way of controlling the immune system so that it would not attack the transplanted heart.

After another 10 years of research, 1980 finally saw the development of immunosuppressive drugs which made transplants a therapeutic possibility. Today, more than 75,000 heart transplants have been performed around the world. Today, nearly 90% of transplant patients survive the first year after surgery and 75% are alive three years later.

We are still in the early years of cell therapy. Remember that the first clinical trials with embryonic SCs started in 2010. We should not expect miracle results and, when success proves elusive, we must also be ready to react, to learn from the setbacks and press ahead with research. We need carefully balanced doses of enthusiasm, humility, perseverance, daring and responsibilities, so that one day we will be able to fulfil the therapeutic promise of stem cells to the fullest.

# Final Note

As stem-cell research is so dynamic, this book would sooner or later become outdated, particularly as regards ongoing clinical trials and their results. I hope that much of the promise of stem cells will be fulfilled in time. Readers can follow this progress on the website:

http://www.ib.usp.br/lance.usp/booksc.

It offers current data on clinical trials with adult and embryonic stem cells, information on the use of iPSC cells to understand the mechanisms involved in various different diseases and to identify new drugs to combat them, and reports on the evolution of tissue-specific SCs for regenerative medicine.

# References

1. KRAUSE, D. S. *et al*. Multi-Organ, Multi-Lineage Engraftment by a Single Bone Marrow-Derived Stem Cell. *Cell* **105** (3): 369–377, 2001.

2. QUAINI, F. *et al*. Chimerism of The Transplanted Heart. *NEJM* **346** (1): 5–15, 2002.

3. MEZEY, E. *et al*. Transplanted Bone Marrow Generates New Neurons in Human Brains. *P.N.A.S.*, v. 100, n. 3, p. 1364–1369, 2003.

4. BELTRAMI, A. P. *et al*. Adult Cardiac Stem Cells Are Multipotent and Support Myocardial Regeneration. *Cell* **114** (6): 763–776, 2003.

5. BOLLI, R. *et al*. Cardiac Stem Cells in Patients with Ischaemic Cardiomyopathy (SCIPIO): Initial Results of a Randomised Phase 1 Trial. *Lancet* **378** (9806): 1847–1857, 2011.

6. MAKKAR, R.R. *et al*. Intracoronary Cardiosphere-Derived Cells for Heart Regeneration after Myocardial Infarction (CADUCEUS): A Prospective, Randomised Phase 1 Trial. *Lancet* **379** (9819): 895–904, 2012.

7. JOHNSON, J. *et al*. Germline Stem Cells and Follicular Renewal in the Postnatal Mammalian Ovary. *Nature*, n. 428, pp. 145–150, 2004.

8. ZOU, K. *et al*. Production of Offspring from a Germline Stem Cell Line Derived from Neonatal Ovaries. *Nature Cell Biol* n. 11, pp. 631–636, 2009.

9. WHITE, Y.A.R. *et al*. Oocyte Formation by Mitotically Active Germ Cells Purified from Ovaries of Reproductive-Age Women. *Nature Med* n. 18, pp. 413–421, 2012.

10. BRINSTER, R. L. *et al*. Germline Transmission of Donor Haplo-Type Following Spermatogonial Transplantation. *PNAS* **91** (24): 11303–11307, 1994.

11. HE, Z. *et al.* Isolation, Characterization, and Culture of Human Spermatogonia. *Biology Reproduct* **82** (2): 363–372, 2010.

12. YANG, S. *et al.* Generation of Haploid Spermatids with Fertilization and Development Capacity from Human Spermatogonial Stem Cells of Cryptorchid Patients. *Stem Cell Rep* **3**: 663–675, 2014.

13. BONNET, D.; DICK, J. E. Human Acute Myeloid Leukemia Is Organized as a Hierarchy that Originates from a Primitive Hematopoietic Cell. *Nature Med* n. 3, pp. 730–737, 1997.

14. AL-HAJJ, M. *et al.* Prospective Identification of Tumorigenic Breast Cancer Cells. *P.N.A.S.* **100** (7): 3983–3988, 2003.

15. SINGH, S. K. *et al.* Identification of Human Brain Tumor Initiating Cells. *Nature*, n. 432, pp. 396–401, 2004.

16. LOFFREDO, F.S. *et al.* Bone Marrow-Derived Cell Therapy Stimulates Endogenous Cardiomyocyte Progenitors and Promotes Cardiac Repair. *Cell Stem Cell* **8** (4) 389–398, 2011.

17. EVANS, M. J.; KAUFMAN, M. H. Establishment in Culture of Pluripotential Cells from Mouse Embryos. *Nature*, n. 292, pp. 154–156, 1981.

18. D'AMOUR, K. A. *et al.* Production of Pancreatic Hormone–Expressing Endocrine Cells from Human Embryonic Stem Cells. *Nature Biotech* **24** (11): 1392–1401, 2006.

19. Thomson , J. A. *et al.* Embryonic Stem Cell Lines Derived from Human Blastocysts. *Science* **282** (5391): 1145–1147, 1998.

20. SCHWARTZ, S. D. *et al.* Human Embryonic Stem Cell-derived Retinal Pigment Epithelium in Patients with Age-related Macular Degeneration and Stargardt's Macular Dystrophy: Follow-up of Two Open-label Phase 1/2 Studies. *Lancet* **385**: 509–516, 2015.

21. CAMPBELL, K. H. S. *et al.* Sheep Cloned by Nuclear Transfer from a Cultured Cell Line. *Nature*, n. 380, pp. 64–66, 1996.

22. TACHIBANA, M. *et al.* Human Embryonic Stem Cells Derived by Somatic Cell Nuclear Transfer. *Cell* **153**: 1–11, 2013.

23. Yamada, M. Human Oocytes Reprogram Adult Somatic Nuclei of a Type 1 Diabetic to Diploid Pluripotent Stem Cells. *Nature* **510**: 533–536, 2014.

24. TAKAHASHI, K. *et al.* Induction of Pluripotent Stem Cells from Mouse Embryonic and Adult Fibroblast Cultures by Defined Factors. *Cell* **126** (4): 663–676, 2006.

25. TAKAHASHI, K. *et al*. Induction of Pluripotent Stem Cells from Adult Human Fibroblasts by Defined Factors. *Cell* **131**(5): 1–12, 2007.

26. VIERBUCHEN, T. *et al*. Direct Conversion of Fibroblasts to Functional Neurons by Defined Factors. *Nature*, n. 463, pp. 1035–1041, 2010.

27. PANG, Z. P. *et al*. Induction of Human Neuronal Cells by Defined Transcription Factors. *Nature*, n. 476, pp. 220–223, 2011.

28. SMART, N. *et al*. *De novo* Cardiomyocytes from within the Activated Adult Heart after Injury. *Nature*, n. 474, pp. 640–644, 2011.

29. HUANG, P. *et al*. Induction of Functional Hepatocyte-Like Cells from Mouse Fibroblasts by Defined Factors. *Nature*, n. 475, pp. 386–389, 2011.

30. SZABO, E. *et al*. Direct Conversion of Human Fibroblasts to Multilineage Blood Progenitors. *Nature*, n. 468, pp. 521–526, 2010.

31. CAIAZZO, M. *et al*. Direct Generation of Functional Dopaminergic Neurons from Mouse and Human Fibroblasts. *Nature*, n. 476, pp. 224–227, 2011.

32. SON, E. Y. *et al*. Conversion of Mouse and Human Fibroblasts into Functional Spinal Motor Neurons. *Cell Stem Cell* **9**(3): 205–218, 2011.

33. DIMOS, J. T. *et al*. Induced Pluripotent Stem Cells Generated from Patients with ALS Can Be Differentiated into Motor Neurons. *Science* **321**(5893): 1218–1221, 2008.

34. ITZHAKI, I. *et al*. Modelling the Long QT Syndrome with Induced Pluripotent Stem Cells. *Nature*, n. 471, pp. 225–229, 2010.

35. AMARIGLIO, N. *et al*. Donor-Derived Brain Tumor Following Neural Stem Cell Transplantation in an Ataxia Telangiectasia Patient. *PLOS Medicine*, Cambridge **6**(2), 2009.

36. Regenberg A.C. *et al*. Medicine on the Fringe: Stem Cell-Based Interventions in Advance of Evidence. *Stem Cells* **27**: 2312–2319, 2009.

The links in this book were accessed in December 2015.

# Index

Made in the USA
San Bernardino, CA
08 May 2018